MANUAL OF OPTICAL MINERALOGY

MANUAL OF
OPTICAL MINERALOGY

DAVID SHELLEY

*Department of Geology, University of Canterbury,
Christchurch, New Zealand*

ELSEVIER SCIENTIFIC PUBLISHING COMPANY
Amsterdam, Oxford, New York 1975

ELSEVIER SCIENTIFIC PUBLISHING COMPANY
335 JAN VAN GALENSTRAAT
P.O. BOX 211, AMSTERDAM, THE NETHERLANDS

AMERICAN ELSEVIER PUBLISHING COMPANY, INC.
52 VANDERBILT AVENUE
NEW YORK, NEW YORK 10017

Library of Congress Cataloging in Publication Data

Shelley, David.
    Manual of optical mineralogy.

    Bibliography: p.
    Includes index.
    1. Optical mineralogy--Handbooks, manuals, etc.
I. Title.
QE369.06S49   549'.125   75-29690

ISBN: 0-444-41303-0 (Hardback)
        0-444-41387-1 (Paperback)

WITH 130 ILLUSTRATIONS AND 25 TABLES

PRINTED IN THE NETHERLANDS

To Iola

Preface

Optical mineralogy forms a major part of most university courses in geology, and is a pre-requisite for petrography. In order to cover all those aspects of crystallography, theory, technique, procedure, and systematics used in the practice of the subject, most teachers find it necessary to recommend several texts. These not only prove expensive to the student, but are often rather too detailed and abstruse for practical work. The prime intention of this book, therefore, is to provide a handy one-volume reference to all the information normally required in the laboratory; it is hoped that the book will be useful as such for all stages of undergraduate and later work.

Many students find difficulty with their microscope work, and are not helped by the many texts in which principles and techniques are almost irretrievably mixed. In this book, the underlying principles are separated from the laboratory techniques; also separately described are the laboratory procedures that should enable the student to develop an efficient but rigorous routine for identifying minerals.

In attempting to explain optical theory it is easy to lose sight of the prime purpose of the exercise, that is to identify minerals successfully. All texts written for mineralogists have simplified theory to some extent, and here it is kept to the bare essentials necessary for understanding the interrelationships of optics and crystallography. Hence, anisotropic minerals are discussed solely in terms of the uniaxial and biaxial indicatrices, no reference being made to wave-front theory.

Chapter 4 explains the standard techniques used to identify minerals in general petrographic work; included is a section on universal-stage methods. The techniques for thin-section work and grain-mount studies are treated separately where they differ significantly.

Each mineral described in Chapter 7 is assigned a number in order of description, and this facilitates rapid cross-reference between the descriptions and Determinative Tables of Chapter 6. The mineralogical data have been brought up to date as far as possible by reference to the literature through *Mineralogical Abstracts*. However, data appertaining to mineralogical oddities which distort the normal ranges of properties have been omitted. In general, the depth of treatment is that considered suitable for undergraduate and routine petrographic work. Information on dispersion in minerals has been omitted since in my experience, very few workers use this property for identification. More orientation diagrams than usual are provided. These are designed to help and encourage the student to check the properties of minerals in several orientations. Except for a few minerals such as the feldspars, information on paragenesis is brief, this being a subject more appropriately dealt with in petrological or theoretical mineralogy texts.

Three more theoretical texts that are thoroughly recommended to students as supplements to this book are *An Introduction to Crystallography* by F.C. Phillips (1971), *Optical Crystallography* by E.E. Wahlstrom (1969), and *An Introduction to the Rock-Forming Minerals* by W.A. Deer, R.A. Howie, and J. Zussman (1966).

Inevitably, a book such as this owes a considerable debt to the many mineralogists and petrographers whose data have been compiled here. Acknowledgement to all is impossible, and is made only where diagrams have been taken directly from their work. I should like to thank Dr. G.J. van der Lingen and Dr. J. Bradshaw for commenting on some parts of the text, and especial thanks go to Lee Leonard for draughting all the figures.

October 1974   DAVID SHELLEY

COPYRIGHT ACKNOWLEDGEMENTS

I am greatly indebted to the authors and the following journals and publishers for permission to reproduce their diagrams as detailed below. Credit to the authors is given in the figure captions.

The *Journal of Petrology* published by the Clarendon Press, Oxford, for Fig. 81. © 1964 Oxford University Press; Fig. 82, © 1963 Oxford University Press; a portion of Fig. 86, © 1963 Oxford University Press; and Fig. 102b, © 1970 Oxford University Press.

The *American Mineralogist* published by the Mineralogical Society of America for Figs. 50, 71, 113, 126, 132, and parts of Figs. 86, 109 and 110.

The *Mineralogical Magazine* published by the Mineralogical Society of Great Britain for Fig. 84 and part of Fig. 86.

The *Journal of Geology* published by the University of Chicago Press for Fig. 83 and part of Fig. 80.

Nature published by MacMillan Journals Ltd. for Fig. 66.

The Geological Society of America for Fig. 102 (a).

The *American Journal of Science* for Fig. 118 and parts of Figs. 103 and 110.

Wiley and Sons Inc. for Fig. 103.

E. Schweizerbart'sche Verlagsbuchhandlung for part of Fig. 109.

Schweizerische Mineralogische und Petrographische Mitteilungen published by Verlag Leeman for Fig. 107.

Abbreviations and Symbols Used in the Text

a, b, c	Crystal axes and cell edges
α, β, γ	Angles between the positive ends of the b and c, a and c, and a and b crystal axes respectively
α, β, γ	Least, intermediate and greatest refractive indices in biaxial crystals
	Note: the particular meaning of α, β and γ in various parts of the text is always obvious or made clear by a note
ω	The ordinary ray or refractive index in uniaxial crystals
ϵ	The extraordinary ray or refractive index in uniaxial crystals
ϵ'	A ray or refractive index in uniaxial crystals intermediate between ω and ϵ
n	Refractive index of isotropic material.
R.I.	Refractive index
δ, δ'	Birefringence, partial birefringence
X, Y, Z	Vibration directions in biaxial crystals of the fastest, intermediate, and slowest rays respectively; equivalent to the refractive indices α, β, and γ
O.A.P.	Optic axial plane
+ve, —ve	Positive, negative
$2V_x, 2V_z$	The optic axial angle about X or Z
$2E$	The apparent optic axial angle in air
C.B.	Canada balsam
λ	Wavelength
ca.	Approximately
//	Parallel to
$\perp$	Perpendicular to
$\wedge$	Angle between two faces or directions
$<, >$	Less than, greater than
(), {}, []	Face, form, and zone symbols enclosing Miller indices (see Chapter 1)

Contents

An Introduction to Crystallography

Crystals may be investigated in a number of ways, the most obvious being to analyse their form and symmetry in hand specimen. More fundamentally perhaps, the atomic structure of crystals can be studied with X-rays. This book is concerned primarily with another branch of crystallography in which the optical properties of crystals are studied. With the aid of the polarising microscope, these properties enable us to identify the common rock-forming minerals quickly and accurately.

Useful as optical properties are, to apply them successfully, it is necessary to understand their relationship to the more general aspects of crystallography. The following account introduces the general crystallographic concepts and terminology required for optical work.

CRYSTALS

A crystal is a solid substance characterised by a regular internal arrangement of its constituent atoms. This internal arrangement can be determined using X-rays. The flat surfaces (*crystal faces*) that bound well-shaped crystals are disposed in such a way that they reflect the internal arrangement of the atoms. Crystals may be distinguished as *euhedral*, if they are bounded by crystal faces, *subhedral*, if they are partially so bounded, or *anhedral*, if no crystal faces are present.

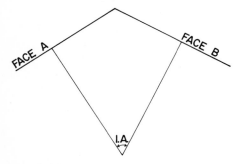

Fig. 1. The interfacial angle (I.A.) to two crystal faces *A* and *B*.

THE INTERFACIAL ANGLE

Two crystals are seldom exactly alike in shape. For various reasons which need not concern us here, certain crystal faces may be better developed in one crystal than in others. Nevertheless, it is always found that the angle between two particular faces for a specific mineral is constant, regardless of how well developed the faces are. This directly reflects the regularity of the internal arrangement of atoms. The angle is expressed as the *interfacial angle*, which is the angle between the normals to the two faces (Fig. 1).

CRYSTAL SYMMETRY

Almost all crystals possess a degree of symmetry, which may be expressed in terms of either axes, planes or a centre of symmetry. When considering the symmetry elements of a crystal, it is important to disregard the imperfections of growth, and the unequal development of faces. Symmetry is judged on the angular relationships between crystal faces.

Axes of symmetry

If the (perfect and regular) crystal has an identical disposition of faces in two or more positions when rotated about a line, that line is termed an axis of symmetry. There are two-fold (diad), three-fold (triad), four-fold (tetrad) and six-fold (hexad) axes of symmetry in various minerals. As examples, a four-fold axis is shown in a simple cube (Fig. 2a), and a six-fold axis in a simple hexagonal prism (Fig. 2b).

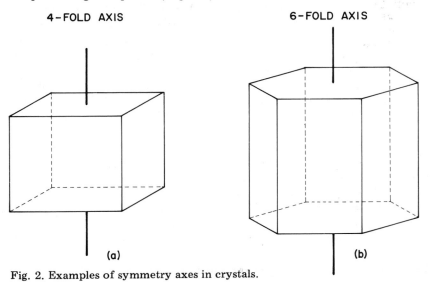

4-FOLD AXIS   6-FOLD AXIS

(a)   (b)

Fig. 2. Examples of symmetry axes in crystals.

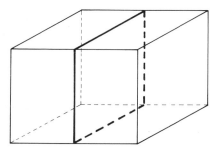

Fig. 3. One of the nine planes of symmetry of a cube.

Planes of symmetry
If the (perfect and regular) crystal can be divided into two identical halves, the plane which so divides it is termed a plane of symmetry. A simple cube has nine such planes of symmetry, and one of these is shown in **Fig. 3**.

Centre of symmetry
If the (perfect and regular) crystal has a central point through which a line in any direction will emerge at an identical point on either side of the crystal, it is said to have a centre of symmetry. A simple cube, for example, clearly has such a centre.

CRYSTAL SYSTEMS

Depending on the particular symmetry elements of a crystal, it may be classified as belonging to one of seven crystal systems. Each crystal system can be described in terms of three or four *crystal axes* known as *a*, *b* or *c*, the choice and disposition of which are a consequence of symmetry. The following crystal systems are normally distinguished (Fig. 4). More detailed information on the symmetry elements possible in each system can be found in standard crystallography texts (Phillips, 1971).

Cubic system. Three crystal axes, a_1, a_2, a_3, all equal and at right angles to each other. All cubic minerals have four triad axes of symmetry.

Tetragonal system. Three crystal axes, two equal, a_1, a_2, and a third, generally unequal, *c*, all at right angles to each other. The *c*-axis is always a tetrad axis of symmetry.

Hexagonal and trigonal systems. Four crystal axes, three equal, a_1, a_2, a_3, which lie in a plane at 120° from each other, and a fourth, generally unequal, *c*, which is at right angles to the plane of the *a*-axes. The *c*-axis is a hexad axis of symmetry in the hexagonal system, and a triad in the trigonal system.

Orthorhombic system. Three crystal axes, *a*, *b*, *c*, in general all unequal, and all at right angles to each other. The c-axis (and often the a- and b-axes) is a diad axis of symmetry.

Monoclinic system. Three crystal axes, a, b, c, in general all unequal. b is at right angles to the ac-plane. The angle between a and c is not a right angle, and the obtuse angle between them is termed β (not to be confused with the refractive index β). The b-axis is always a diad axis of symmetry.

Triclinic system. Three crystal axes, a, b, c, in general all unequal, and none of them at right angles to each other. The angles between the positive ends (Fig. 4) of b and c, a and c, and a and b, are termed α, β and γ respectively (not to be confused with the refractive indices α, β and γ). Only a centre of symmetry is possible.

UNIT CELLS AND AXIAL RATIOS

The basic atomic structure which is repeated regularly in a crystal is termed the *unit cell*. X-ray studies enable the size of the unit cell to be determined. Its dimensions are expressed in terms of lengths along the direction of the crystal axes a, b, and c. The *axial ratio* simplifies these absolute lengths into relative values. For tetragonal, hexagonal, or trigonal minerals, the axial ratio is expressed in terms of the ratio c : a, a being taken as 1. For orthorhombic, monoclinic, or triclinic minerals, the axial ratio is expressed in terms of the ratio a : b : c, b always being taken as 1.

MILLER INDICES

There are several ways in which the angular relationship of a crystal face to the crystal axes can be expressed. The most common method is by means of Miller indices. These denote the reciprocals of the distances by which a crystal face intercepts the crystal axes, the distances being measured in units proportional to the axial ratio. Three indices (or four for the hexagonal and trigonal systems) are given, one for each of the crystal axes, and they are always expressed as whole numbers or zero. If a face is parallel to a crystal axis, it intercepts it at infinity, the reciprocal of which is zero. If a face intercepts the negative end of a crystal axis, a bar is placed over the number. A few examples will serve to clarify their use.

Cube (Fig. 5a)

Face A cuts a_1, but is parallel to a_2 and a_3. The reciprocals of the intercept distances are therefore a_1 (1), a_2 ($\frac{1}{\infty} = 0$), a_3 ($\frac{1}{\infty} = 0$). The Miller indices are (100). Similarly, face B is (010) and face C (001).

Pyritohedron (Fig. 5b)

Face A is parallel to a_3, and cuts a_1 at half the distance from the origin as

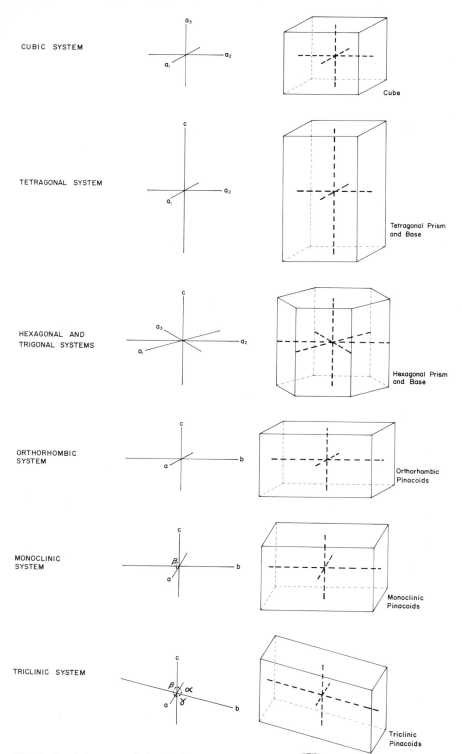

CUBIC SYSTEM

Cube

TETRAGONAL SYSTEM

Tetragonal Prism and Base

HEXAGONAL AND TRIGONAL SYSTEMS

Hexagonal Prism and Base

ORTHORHOMBIC SYSTEM

Orthorhombic Pinacoids

MONOCLINIC SYSTEM

Monoclinic Pinacoids

TRICLINIC SYSTEM

Triclinic Pinacoids

Fig. 4. Crystal axes and simple forms of the crystal systems.

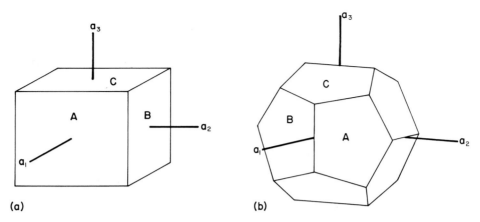

Fig. 5. The text explains how to derive Miller indices for the crystal faces A, B and C of a cube (a), and pyritohedron (b).

a_2. The relative intercept distances are a_1 (1), a_2 (2), a_3 (∞), and the reciprocals are $\frac{1}{1}$, $\frac{1}{2}$, $\frac{1}{\infty}$, which expressed as whole numbers (Miller indices) is (210). Face B intercepts at the relative distances of 1, $\bar{2}$, and ∞, and the Miller indices are ($2\bar{1}0$). Face C intercepts at the relative distances 2, ∞ and 1, and the Miller indices are (102).

Tetragonal pyramid (Fig. 6)

Face A is parallel to a_2 and intercepts the c and a_1 axes at unequal

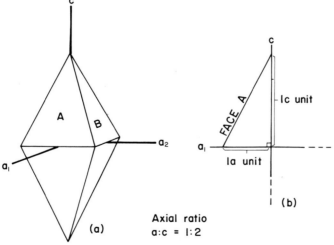

Fig. 6. The text explains how to derive Miller indices for the faces A and B of a tetragonal pyramid.

distances. However, these distances are in the same relative proportion as the axial lengths as expressed in the ratio $c : a = 2 : 1$ (Fig. 6b). The Miller indices are therefore (101). Face B is parallel to a_1 but intercepts c and a_2 at distances in the same proportion as the axial lengths, and is therefore (011).

Monoclinic feldspar crystal (Fig. 7)

Face A intercepts a and c at ∞, and the Miller indices are (010). Face B intercepts a and b at distances directly proportional to the unit lengths, and is parallel to c. The Miller indices are therefore (110). Face C is identical to B except that it cuts the negative end of b, and is therefore $(1\bar{1}0)$. Face D is parallel to both b and a (Fig. 7b), and is therefore (001). Face E is parallel to b, but cuts a and c. The intercept distance along a is half the unit length compared to its intercept along c (Fig. 7b). Therefore the relative intercept lengths can be expressed as 1, ∞ and $\bar{2}$, which as the reciprocal Miller indices is $(20\bar{1})$.

Note: The Miller indices of individual crystal faces are always placed in ordinary brackets, e.g. (101). Miller indices may be more generally expressed as (hkl) of (hkil), where h + k + i always equals zero.

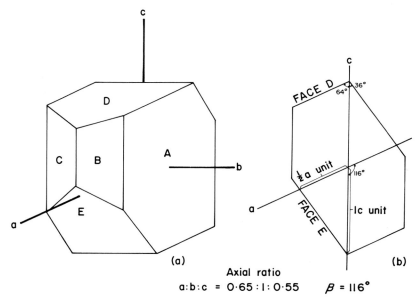

Fig. 7. The text explains how to derive Miller indices for the faces A, B, C, D and E of a monoclinic feldspar crystal.

CRYSTAL FORMS

A crystal form is a face or a group of faces which have identical relationships to the crystal axes by virtue of the crystal symmetry. The form can be expressed in terms of the Miller indices of one representative face, the Miller indices being placed in braces, e.g. $\{100\}$.

A cube form, for example is denoted by $\{100\}$ which refers not only to (100), but also to (010), (001), $(\bar{1}00)$, $(0\bar{1}0)$ and $(00\bar{1})$, since all these faces are identical with respect to symmetry. A form is *closed* if it can completely enclose space and exist by itself. *Open* forms do not enclose space, and must exist in combination with other forms.

Some form names in common use are described below.

Cube $\{100\}$. The closed form comprising the six sides of the simple cube, all of which are identical in terms of symmetry.

Other cubic-system forms. There are several forms in the cubic system, each of which has several faces, all identical in terms of symmetry. These include the octahedron $\{111\}$, the rhombdodecahedron $\{110\}$, the tetrahexahedron $\{hk0\}$, and the trapezohedron $\{hll\}$, all of which are closed forms.

For crystals other than cubic, the following are common forms.

Pyramid. A form comprising several nonparallel faces that meet at a point, e.g. $\{111\}$ in the tetragonal system.

Rhombohedron. A closed form comprising six faces whose intersection edges are not at right angles, e.g. $\{10\bar{1}1\}$.

Prism. An open form of several faces all of which are parallel to the same axis; this axis is most often c. Common form indices are $\{110\}$ and $\{10\bar{1}0\}$.

Pinacoid. An open form comprising two parallel faces, e.g. $\{010\}$ in the monoclinic system.

CRYSTAL ZONES

A zone is a group of crystal faces which intersect in a set of parallel edges. The direction of the parallel edges is known as the zone axis, and is expressed as coordinates derived from Miller indices, and placed in square brackets $[uvw]$. If the Miller indices of two faces $(h_1k_1l_1)$ and $(h_2k_2l_2)$ in a zone are known, then the zone axis is derived as follows:

$$u = k_1l_2 - l_1k_2$$

$$v = l_1h_2 - h_1l_2$$

$$w = h_1k_2 - k_1h_2$$

For example the faces (210) and (110) belong to a zone defined by an axis [001]. Zone axes are commonly cited instead of crystal axes. Thus [001] may be used instead of the c-axis.

CRYSTAL HABIT

Crystals, even of one species, vary considerably in shape, depending on rates of growths, impurities present during growth, and a host of other factors. Nevertheless, particular species are characterised by certain shapes called crystal habits. Well-known terms that are self explanatory are *fibrous*, *acicular* (needle-like), *columnar*, *tabular*, *scaly*, *micaceous*. In addition, form names are used if a particular form is well developed, hence *cubic*, *prismatic*, *pyramidal*, etc.

CLEAVAGE, FRACTURE AND PARTING

Many crystals break easily along smooth planes which are parallel to possible crystal faces, usually simple index ones. Such planes are termed cleavage planes. Cleavages are repeated by the symmetry of a crystal in exactly the same way as faces. It will be found in the systematic description of minerals, that the indices of a cleavage are often placed in braces like crystal forms. Hence in a cubic mineral, $\{100\}$ cleavage refers not only to (100), but also to the (010) and (001) cleavages, all of which are identical in terms of symmetry. On the other hand, in a triclinic mineral with little or no symmetry, $\{100\}$ refers to a single plane (100) since there are no other planes identical in terms of symmetry.

A cleavage may be described as *perfect*, *good*, *distinct*, *imperfect*, or *poor*, depending on its ease of development.

Fracture refers to the shape of surfaces formed by breaking a crystal along directions other than cleavages. If a crystal has a number of perfect cleavages, fracture may be difficult to observe. Fracture may be *conchoidal* (shell-like surfaces), *even* (sub-planar), *uneven*, or *hackly* (jagged).

Parting refers to planes of separation which bound twin planes or exsolution lamellae in a crystal. It may be confused with cleavage.

TWINNED CRYSTALS

A twinned crystal is formed of two or more individuals of the same species, joined together according to a definite law. They may be joined together as *contact twins*, simply united by a common plane, or as *penetration twins*, where they appear to cross each other in a complex, but symmetrical way. *Simple twins* consist of just two individuals. Repeated or *polysynthetic twins* consist of several individuals, often in lamellar form.

The geometric relationship between the individuals of a *twinned crystal* can be described in terms of *twin axes* or *twin planes*. A twin axis is a line of rotation about which one twin can be brought into the orientation of the other (usually by a rotation of 180°). A twin axis is the most convenient way of describing most twins. A twin plane is a plane of reflection across which the twins are mirror images. Obviously, a twin plane cannot be parallel to a plane of symmetry of the crystal since the mirror images across such a plane are identical. The twin plane must not be confused with the *composition plane* which is the plane that actually unites the two individuals, though in some cases the two planes coincide.

Normal twins have a twin axis normal to the composition plane. *Parallel twins* have a twin axis that lies in the composition plane parallel to a crystal edge. *Complex twins* can be visualised as a combination of normal and parallel-twinning, the twin axis lying in the composition plane normal to a crystal edge (for examples refer to the plagioclase-feldspars).

Twin crystals may form in several ways. *Growth* or *primary twins* form during the growth of the crystal, often at the beginning of growth. *Secondary twins* form subsequent to crystal growth. For example, *deformation twins* result from the rotation of part of the crystal into a twin orientation during deformation. Calcite develops deformation twins very easily, if squeezed in a vice. *Transformation twins* result from the change of symmetry of certain crystals on changes of temperature and pressure. For example, monoclinic alkali-feldspar may invert to a triclinic form on cooling, and in so doing will nucleate numerous twins in a cross-hatched arrangement (see the section on feldspars). Leucite twins in a similar way on changing from cubic to a lower symmetry on cooling.

CHAPTER 2

The Polarising Microscope

The polarising microscope (also known as the petrological or petrographic microscope) is the principal piece of equipment used by the geologist to observe the optical properties of minerals. There are numerous microscopes available on the market today and they vary considerably in their construction details. This applies particularly to the illumination and sub-stage condenser systems, some of which are designed for routine, uncritical work, whilst others require or allow detailed adjustments to be made. The student will have to learn how to correctly adjust these systems for the particular microscope he is provided with, and he should read the appropriate microscope manual.

The purpose of this chapter is to familiarise the student with the terminology and function of the various parts of the microscope. A Zeiss microscope (1970 — RP 48 model) and a Swift microscope (1961 model) are used for illustration (Figs. 8 and 9). The detailed manipulation of the microscope for observing particular optical properties is described in the appropriate parts of the succeeding three chapters.

THE ROTATING STAGE

All petrological microscopes are fitted with a rotating stage which is graduated in degrees. A vernier scale is usually fitted adjacent to the stage so that tenths of a degree of rotation can be measured. A clamp enables the stage to be fixed in any position.

THE POLARISER AND ANALYSER

In modern microscopes, sheets of polaroid are used to produce polarised light (q.v.). One sheet, called the polariser, is placed below the stage, and on the Zeiss and Swift microscopes allows light to pass through it vibrating E—W (parallel to the E—W cross-hair of the eyepiece). Some other microscopes (e.g. Leitz and Nikon) have the polariser orientated with a N—S vibration direction. The polariser is normally left fixed in position, but on most microscopes it can be removed or rotated if necessary.

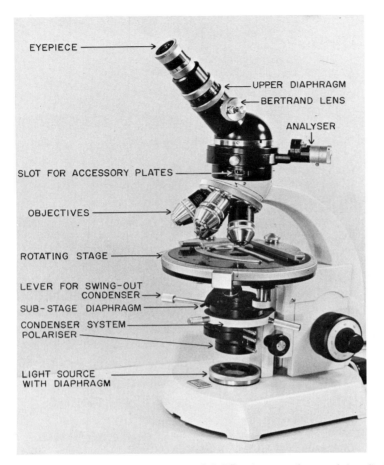

EYEPIECE

UPPER DIAPHRAGM

BERTRAND LENS

ANALYSER

SLOT FOR ACCESSORY PLATES

OBJECTIVES

ROTATING STAGE

LEVER FOR SWING-OUT
                 CONDENSER

SUB-STAGE DIAPHRAGM

CONDENSER SYSTEM
POLARISER

LIGHT SOURCE
WITH DIAPHRAGM

Fig. 8. Polarising microscope model RP 48, manufactured by Carl Zeiss, W. Germany (1970).

The analyser is a second sheet of polaroid placed between the objective and the eyepiece. It allows light to vibrate in a direction (N—S on the Zeiss and Swift) at right angles to the polariser vibration direction. It is designed to be either swung in or out of the beam of light, depending on the particular optical properties to be observed. On some more sophisticated microscopes (e.g. the Zeiss) the analyser can also be rotated.

The student should periodically check: (1) that the polariser and analyser vibration directions are at right angles to each other, and (2) that their vibration directions are orientated parallel to the cross-hairs of the eyepiece. This is easily done as follows: (1) no light should reach the eye when both polariser and analyser are placed in the beam of light (with no mineral

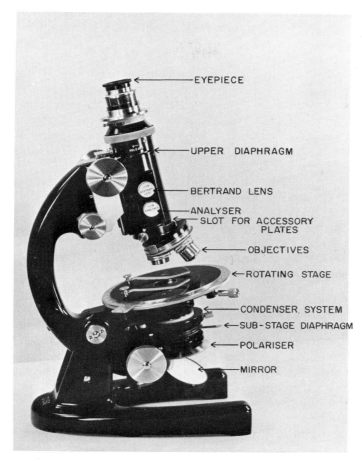

Fig. 9. Polarising microscope manufactured by James Swift and Sons Ltd., England (1961).

specimen on the stage); and (2) prismatic grains of a mineral with straight extinction (q.v.) such as sillimanite, should extinguish when the prismatic edges are parallel to the cross-hairs.

THE OBJECTIVES

These are the lenses used for magnifying the specimen on the stage. Three, four or five are normally supplied as standard equipment and include objectives of low power ($\times 2$—$\times 4$) for viewing large fields of view, of medium power ($\times 10$) for general optical work, and of high power ($\times 40$—$\times 50$) for observing interference figures and very small grains. On most modern microscopes, the objectives are fitted to a rotating nose-piece which allows quick and easy change of magnification.

The *depth of focus* of a particular objective refers to the vertical distance that can be seen simultaneously in focus with the objective. It may be increased by closing the sub-stage diaphragm. The *working distance* is the distance between the objective lens and the specimen, when the specimen is in focus. The *numerical aperture* (N.A.) is a measure of the cone of light that enters the objective lens when in focus. N.A. is equal to $n \sin \mu$ where n is the refractive index of the medium between the specimen and the objective ($n = 1$ for air), and μ is the half angle of the cone of light.

There are several types of objective designed to overcome various optical aberrations. Chromatic aberration is caused by the separate resolution of the colours of the spectrum resulting in colour fringes. Spherical aberration results in the image being focussed in different planes for light passing through the margins and centre of the lens. *Achromatic* objectives correct for the chromatic aberration. *Planachromats* also correct for the spherical aberration, and are preferable, especially for low magnifications.

THE EYEPIECE (OR OCULAR)

The eyepiece is the lens fitted to the top of the microscope tube, and it magnifies and focusses the image produced by the objective lens. The eyepiece assembly contains two cross-hairs, and is slotted into the microscope tube so that the cross-hairs are orientated E—W and N—S, i.e. parallel to the vibration directions of the polariser and analyser. The lens of the eyepiece can be raised or lowered in its mount so that the cross-hairs are focussed.

Most eyepieces have a magnification of $\times 8$ or $\times 10$. The magnification produced by both eyepiece and objective is obtained by multiplying the two separate magnifications. The student should note the sizes of the fields of view for various combinations of eyepiece and objective. These can be measured using a stage micrometer.

CONDENSER SYSTEM

The condenser consists of a number of lenses positioned below the stage, and which can be adjusted to vary the beam of light impinging on the specimen. The condenser systems incorporated into different microscopes vary considerably, and the student should refer to the appropriate microscope manual for details.

For ordinary viewing, the focal plane of the condenser system should coincide with the focal plane of the objective. This is achieved by raising or lowering the condenser until the maximum resolution is obtained.

On some microscopes (including the Zeiss model of Fig. 8), there is an additional swing-out condenser, used only when observing interference figures. It should be used with the ordinary condenser system fully raised.

SUB-STAGE DIAPHRAGM

An iris diaphragm is incorporated in most condenser systems, and should be adjusted for ordinary viewing as detailed in the following section on illumination. For observing relief and Becke lines, it is usually necessary to partially close this diaphragm.

ILLUMINATION SYSTEM AND MIRROR

Illumination may be by daylight, but normally some form of electric lamp is used. If the light source is an ordinary tungsten-filament bulb, a blue filter should be used to reduce the yellow colour of the light. A ground-glass plate may be necessary to diffuse the light from high intensity bulbs and prevent an image of the light filament being seen. Some microscopes such as the Zeiss have built-in illuminators, but others such as the Swift use a mirror to direct light along the microscope tube.

Most microscopes require some adjustments to be made to the illumination system depending on the type of observation to be made. If a mirror is used, the first adjustment should be to ensure that is is tilted so as to reflect light directly along the microscope tube (many students forget to do this!). The mirror has on one side a plane surface which is normally used; the other concave surface is sometimes used to improve the illumination. For ordinary viewing the important requirement is to adjust the illuminator and sub-stage diaphragm until an evenly illuminated field of view is obtained. If critical illumination is required, the following adjustments should be made: the diaphragm in the light source should be adjusted so that for every objective used, the field of view is *just* illuminated; the diaphragm of the sub-stage condenser system should be adjusted for each objective by closing it until the illumination of the field of view just begins to dim.

THE BERTRAND LENS

The Bertrand lens magnifies and focusses interference figures, and is swung into position in the microscope tube when so required. An alternative means of viewing interference figures is to remove the eyepiece and look down the microscope tube at the high-power objective lens, preferably with the aid of a pin-hole stop inserted at the top of the tube.

UPPER DIAPHRAGM

Both the Swift and Zeiss microscopes have an iris diaphragm in the upper part of the microscope tube. It is used in conjunction with the Bertrand lens, and should be partially closed if interference figures from very small grains are to be observed. Many other microscopes incorporate a fixed diaphragm in the Bertrand lens, and adjustment is not possible.

ACCESSORY PLATES

Accessory plates consist of mineral sections of a thickness such that they produce a known amount of retardation (q.v.). They are used for studying interference figures and the retardation produced by mineral specimens. When required, they are inserted into the microscope tube in a slot between the objectives and the analyser.

CENTRING THE MICROSCOPE

When the stage is rotated, the axis of rotation should coincide with the centre of the field of view. This is achieved on the Zeiss microscope by adjusting a collar on the barrel of each objective. The Swift is adjusted by means of two screws set into the microscope tube just above the objectives. Some other microscopes have centring screws fitted to the stage. Unless each objective is centred independently, as with the Zeiss, it is important to ensure that the centring is most precise for the highest-power objective normally used. To centre a microscope, the point about which an object is seen to rotate when the stage is rotated must be brought to the centre of the cross-hairs by adjusting the centring screws. If the microscope is badly out of centre, centre first for low power, then more precisely for high power.

CARE OF THE MICROSCOPE

The microscope should be protected from dust as much as possible. The lenses should be cleaned with special lens tissue, or with a soft brush, preferably one attached to a small bellows for blowing away coarse material. If a liquid cleaner is necessary, use xylol or benzene but not alcohol which dissolves the cements in some objectives.

The microscope should always be carried by holding the main-arm support, not by holding the microscope tube.

When focussing on specimens using high power, the working distance is so small that special care must be exercised. The objective should be lowered until it almost touches the specimen (observed from the side), and then slowly raised until brought into focus. When viewing with the high-power objective use only the fine-focussing adjustment.

CHAPTER 3

Principles of Optical Mineralogy

Most texts integrate the discussion of optical theory with a description of laboratory techniques. For easier laboratory reference, this book separates as far as possible the background theory from its practical application. Thus, the present chapter discusses in principle the nature of the three most important optical properties of minerals, viz. refraction and refractive index, birefringence, and the optical indicatrix. The succeeding chapter describes how these properties are measured in practice, and how they may be interrelated with such properties as cleavage and colour.

All books on optical mineralogy simplify optical theory to some extent, and this book is no exception. This chapter provides what, in my experience, is just enough theory for successful laboratory work in identifying minerals.

POLARISED LIGHT

The optical properties of minerals are most simply explained in terms of the electro-magnetic theory of light. According to this theory, light may be discussed in terms of wavelengths (Fig. 10a) and vibrations perpendicular to the direction of propagation. White light comprises a gradational series of wavelengths from 390 nm (violet) to 770 nm (red), and may be separated into its component colours by the well-known triangular glass prism. The vibrations of ordinary light are considered to take place in *all* directions perpendicular to the propagation direction (Fig. 10b). Polarised light vibrates in *only one* direction perpendicular to the propagation direction (Fig. 10c).

Polarised light is produced by the polariser and analyser, both of which in modern microscopes consist of a sheet of plastic (polaroid) which absorbs all light except that vibrating in one direction. Natural tourmaline crystals also strongly absorb light vibrating in all but one direction, and may be used as simple polarisers. Older microscopes employed an ingeneous combination of calcite prisms to produce polarised light (described first by William Nicol in 1829, and known as *Nicol prisms*).

Observations with the petrographic microscope may be:

(1) *In ordinary light.* Such a system is seldom used.

(2) *In plane-polarised light.* Light is polarised by the polariser (below the stage) before passing through the specimen on the stage.

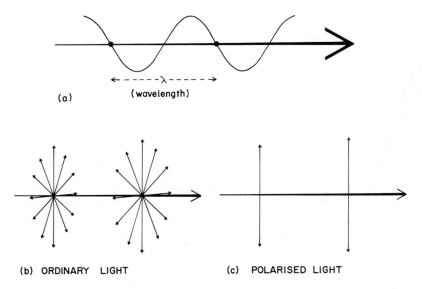

(a)   (wavelength)

(b) ORDINARY LIGHT   (c) POLARISED LIGHT

Fig. 10. Illustrating the concepts of: (a) wavelengths of light, (b) vibration directions of ordinary light, and (c) the single vibration direction of polarised light.

(3) *In crossed-polarised light*. In addition to the sub-stage polariser the analyser above the microscope stage is inserted. Light is then polarised after, as well as before passing through the specimen. The two sheets of polaroid are oriented so that they transmit light vibrating in directions at right angles to each other (E—W and N—S). The effect of superimposing one upon the other is to absorb all light passing through the microscope (Fig. 11). How-

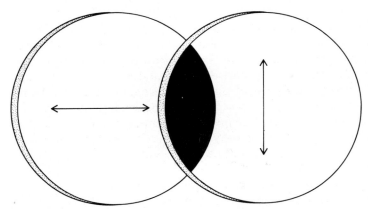

Fig. 11. Two sheets of polaroid with mutually perpendicular vibration directions absorb all transmitted light when superimposed.

ever, many minerals, when placed between the two sheets of polaroid so oriented, affect the light so that it once more reaches the eye.

ISOTROPIC AND ANISOTROPIC MINERALS

The majority of minerals polarise the light that passes through them. In general, two rays of light are produced; their vibration directions are both at right angles to the direction of propagation of the light and at right angles to each other (this is not strictly true, but such an assumption simplifies the following explanations without detracting from the essential principles). Minerals that affect light in such a way are termed *anisotropic*. All crystals except those that belong to the cubic system are anisotropic (though a few anisotropic minerals may in practice appear to be isotropic, e.g. perovskite).

Cubic minerals do not polarise light passing through them, and they do not vary directionally in their effect on light. Cubic minerals are therefore termed *isotropic*. Glass and amorphous substances are also isotropic.

Isotropism directly reflects the high degree of regularity in the atomic structure of cubic minerals. The specific details of anisotropic optical properties also reflect the particular symmetry of the crystals, as will be discussed later.

REFRACTIVE INDEX AND THE VELOCITY OF LIGHT

A ray of light is usually bent when passing from one substance to another (Fig. 12), a phenomenon known as refraction. The refractive index of a substance is given by the ratio of the sines of the angles of incidence and

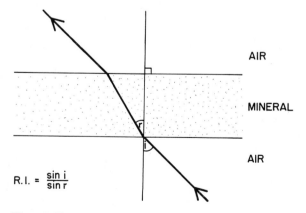

$$R.I. = \frac{\sin i}{\sin r}$$

Fig. 12. Illustrating the phenomenon of refraction.

refraction of light passing from air into the substance (the refractive index of air is taken to be 1).

It can be shown simply that the refractive index of a substance is equal to the ratio between the velocity of light in air and in the substance. In other words, the higher the refractive index of a substance, the slower the light is in passing through it.

The refractive index of common minerals ranges from 1.43 to 3.22, and methods for its determination are described in the chapter on techniques.

Isotropic minerals can be given a single value of refractive index. Anisotropic minerals, however, display the phenomenon of double refraction.

Double refraction

If a small clear cleavage rhomb of calcite is placed over a small dot, two images of the dot will be seen. The two images represent the two separately polarised rays of light produced by the anisotropic calcite. The fact that two images are seen demonstrates that each ray represents a different refractive index for the crystal (and also a different velocity of light). This phenomenon is known as *double refraction*, and is characteristic of all anisotropic minerals, although the amount of double refraction is unusually large in calcite. The refractive index of all anisotropic minerals varies continuously depending on the vibration direction of the light within the crystal.

ISOTROPIC MINERALS IN CROSSED-POLARISED LIGHT

Light is not polarised by cubic minerals. Therefore when light travels through such a mineral placed on the microscope stage, the vibration direction of light produced by the polariser is not changed (Fig. 13), regardless of how we rotate the stage. The light that reaches the analyser is thus polarised at right angles to the vibration direction of the analyser (Fig. 13), and is completely absorbed. In other words, all cubic minerals (and glass) will appear black whenever observed in crossed-polarised light. Only special sections of anisotropic minerals have this property.

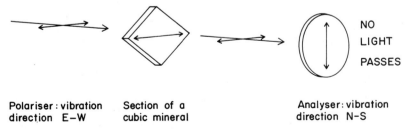

Polariser: vibration   Section of a   Analyser: vibration
direction E—W   cubic mineral   direction N—S

Fig. 13. Behaviour of isotropic minerals in crossed-polarised light.

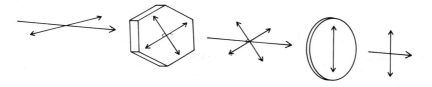

Polariser : vibration	Section of	Analyser : vibration
direction E–W	anistropic mineral	direction N-S

Fig. 14. Anisotropic mineral section in crossed-polarised light with its vibration directions not parallel to those of the polariser and analyser.

ANISOTROPIC MINERALS IN CROSSED-POLARISED LIGHT

Anisotropic minerals generally polarise light into two rays vibrating at right angles to each other. In the general situation, where these vibration directions are not parallel to those of the polariser and analyser, light from the polariser is resolved into two rays by the mineral (Fig. 14). These two rays are resolved by the analyser so that the light reaching the eye is only vibrating in the one direction allowed by the analyser.

If the stage is rotated through 360°, there are four positions in which the two vibration directions of the mineral are parallel to those of the polariser and analyser (Fig. 15). In these positions, light from the polariser cannot be resolved into two rays, and passes through the mineral unchanged in vibration direction. This light is vibrating at right angles to the direction allowed by the analyser, which completely absorbs it. In general therefore, all aniso-tropic minerals will go black (or will *extinguish*) in these four positions when the stage is rotated through 360°.

As we shall see later, there are special sections of anisotropic minerals that remain black or very dark when the stage is rotated, and it is then necessary to make further tests to distinguish them from isotropic mineral sections.

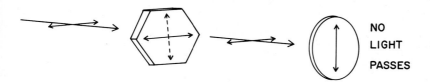

Polariser : vibration	Mineral section	Analyser : vibration
direction E–W		direction N-S

Fig. 15. Anisotropic mineral section in crossed-polarised light with its vibration directions parallel to those of the polariser and analyser.

Birefringence and interference colours

We have already seen that the refractive index of anisotropic minerals varies with direction, and that light passing through them is resolved into two rays, each representing a different refractive index. The maximum possible difference in refractive indices of a mineral is called the *birefringence*. This is an important property that can be readily measured in most thin sections by observing interference colours. Details of the technique are given in Chapter 4, but we shall consider here the principles governing the production of interference colours.

Let us first consider light of one wavelength (monochromatic light, for example sodium light with a wavelength of 589 nm) passing through a mineral section. The light is resolved into two rays which are refracted separately and vibrate in two planes perpendicular to each other (Fig. 16); the rays also differ in wavelength (λ decreases with increase in R.I.). Because of the differences in wavelength (which can also be discussed in terms of distance travelled or speed), each ray completes a different number of wavelengths in the section. Thus in Fig. 16a, ray A completes 2 and ray B $2\frac{1}{2}$ wavelengths. When these two rays are resolved by the analyser to vibrate in one plane they

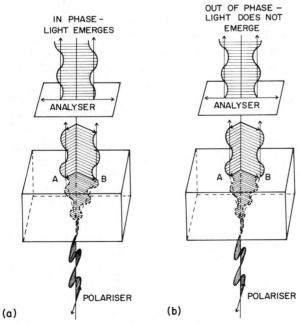

Fig. 16. Illustration showing the effects of two differing retardations produced by two anisotropic mineral sections of differing birefringence in crossed-polarised light (monochromatic). See text for details.

are found to be in phase (Fig. 16a), and reinforce each other so that light is seen. In Fig. 16b, however, the double refraction is increased so that ray *B* completes 3 wavelengths compared to the 2 of ray *A*. They are found to be out of phase with each other when resolved by the analyser, at which time they interfere, cancel each other out, and no light emerges. The number of wavelengths by which ray *B* falls behind ray *A* is called the *retardation*, and the amount of retardation will increase either with an increase in double refraction or an increase in thickness of the mineral section.

The situation with white light is more complex since it consists of a spectrum of colours of gradational wavelengths. After passing through an anisotropic mineral section between crossed-polars, certain colours will be out of phase with each other while others are in phase. Therefore, instead of white light, we see what is known as an *interference colour* that results from the subtraction of the "out-of-phase colours" from the complete spectrum*. The actual interference colour produced changes according to the amount of retardation, and the range of colours commonly seen in thin section are illustrated on the Michel-Lévy chart (Plate 1).

Referring back to Fig. 16, it has already been shown that the retardation (and hence the interference colour) depends on two factors: (1) the thickness of the section (the thicker the section the greater the retardation); and (2) the amount of double refraction. The amount of double refraction depends in turn on two factors: (1) the birefringence of the mineral; and (2) the orientation of the mineral section with respect to the crystal lattice. In order to understand this dependence on orientation of the section, we have to look in further detail at the variation of refractive index within anisotropic minerals.

There are two main groups of anisotropic minerals — uniaxial and biaxial. There is a close relationship between these two groups and crystallography, uniaxial minerals belonging to the hexagonal, trigonal or tetragonal systems, biaxial minerals belonging to the orthorhombic, monoclinic or triclinic systems.

Uniaxial minerals

Hexagonal, trigonal and tetragonal minerals are all characterised by two or three equal crystal axes (*a*-axes) that lie in a plane at right angles to an axis (*c*-axis) of different length (see Chapter 1). The refractive indices within such minerals reflect this characteristic, and are equal for light vibrating in all directions in the plane of the *a*-axes, but vary towards a maximum or minimum value for light vibrating parallel to the *c*-axis. This refractive-index

---

* The student may question the need for the polariser below the stage to observe interference colours and some of the other properties described in this chapter. The purpose of the polariser can be understood in terms of a vector analysis of light as explained in more comprehensive theoretical texts such as Wahlstrom (1969).

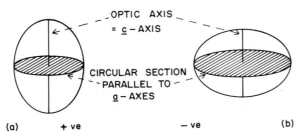

Fig. 17. The uniaxial indicatrix.

variation can be represented by what is termed the uniaxial indicatrix (Fig. 17).

The indicatrix is an ellipsoid whose radii are directly proportional in length to the refractive indices of the mineral for light vibrating in those directions. It is most important to understand from the outset that the indicatrix represents the R.I. of the mineral for vibration directions, and not for travel directions of light. Some uniaxial minerals have a maximum refractive index for light vibrating parallel to the c-axis and these are called uniaxial positive (Fig. 17a). Those with a minimum refractive index for light vibrating parallel to the c-axis are called uniaxial negative (Fig. 17b).

There is an important rule governing the vibration directions of the two light rays passing through any section of a uniaxial mineral. We already know that the two vibration directions are at right angles to each other and at right angles to the direction of propagation (the microscope tube). It is also found that one of the vibration directions always lies in the circular cross-section of the indicatrix, that is, in the plane of the a-axes of the crystal. The light ray with this vibration direction clearly represents a constant value of refractive index for any one mineral, and is therefore called the *ordinary ray* (abbreviated ω). The other ray represents a variable refractive index, which reaches a maximum (uniaxial positive) or minimum (uniaxial negative) value when light vibrates parallel to the c-axis. It is therefore called the *extraordinary ray* (abbreviated ϵ).

To illustrate these points more clearly, let us examine three types of section of uniaxial minerals which will be encountered in laboratory work. Uniaxial +ve quartz is used as an example.

(1) *With the c-axis parallel to the mineral section* (Fig. 18). In this case, light is polarised into the following two rays: one (ω) which has a vibration direction along the line where the circular section of the indicatrix intersects the plane of the mineral section; another (ϵ) which has a vibration direction at right angles to ω and parallel to the mineral section, and in this special case parallel to the c-axis thus representing the maximum R.I. possible (in a +ve mineral). The double refraction in this section is the maximum possible and defines the birefringence of the mineral.

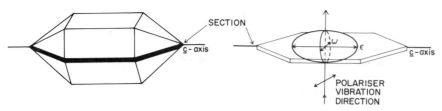

Fig. 18. The uniaxial indicatrix and vibration directions in quartz cut parallel to the c-axis.

(2) *With the c-axis at some angle other than 90° to the mineral section* (Fig. 19). In this case, light is polarised into the following two rays: one (ω) which has a vibration direction along the line where the circular section of the indicatrix intersects the plane of the mineral section; another (ϵ') which has a vibration direction at right angles to ω, but which also lies in the mineral section, and represents an R.I. value intermediate between ϵ and ω. Only a *partial birefringence* is displayed by this section.

(3) *With the c-axis perpendicular to the mineral section* (Fig. 20). In this special case, light can only vibrate in the circular section of the indicatrix, and all possible vibration directions represent equal refractive indices. The result is that light is not doubly refracted, and passes through the mineral with the vibration direction of the polariser unchanged. Inserting the analyser causes total absorption of the light, and the section will appear black in any position of rotation of the microscope stage. In other words the partial birefringence of this section is zero. The unique direction perpendicular to this section and parallel to the c-axis is called the *optic axis*. The fact that there is only one such direction gives rise to the name uniaxial. A uniaxial mineral section with a zero partial birefringence can be distinguished from isotropic mineral sections by means of interference figures (see later).

One further aspect of routine terminology can be introduced at this point. As an alternative to discussing optical properties of anisotropic minerals in terms of relative values of R.I., we can talk in terms of the relative speeds of light rays. Thus in a positive uniaxial mineral, ω is the faster ray and ϵ the slower ray. In a negative uniaxial mineral, ω is the slower, ϵ the faster ray.

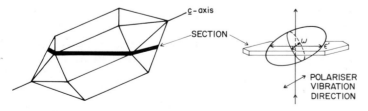

Fig. 19. The uniaxial indicatrix and vibration directions in quartz cut oblique to the c-axis.

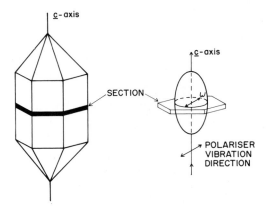

Fig. 20. The uniaxial indicatrix and single vibration direction in quartz cut perpendicular to the c-axis.

Biaxial minerals

Orthorhombic, monoclinic and triclinic minerals are all characterised by three unequal crystal axes (Chapter 1). The refractive indices within such minerals reflect this characteristic, and can be discussed in terms of three unequal values in three principal directions. Unlike the crystal axes, the three optical directions are always at right angles to each other.

We can represent the three optical directions by means of the biaxial indicatrix, which is an ellipsoid with three unequal principal radii (Fig. 21). The shortest and longest radii are constructed parallel to the vibration directions of light representing the smallest and largest refractive indices (or the fastest and slowest rays respectively). These vibration directions are termed X and Z respectively. The intermediate radius at right angles to both X and Z is termed Y. As with the uniaxial indicatrix, it is essential to understand that the directions, X, Y and Z of the indicatrix are proportional in length to the refractive index of light vibrating in those directions. The light rays actually travel at right angles to their vibration directions.

Rays of light passing through biaxial minerals do not necessarily vibrate in any of the directions X, Y and Z, and all three directions are therefore extraordinary. The refractive indices of the crystal when light vibrates parallel to X, Y or Z are termed α, β and γ respectively.

There are the following relationships between crystal axes and X, Y and Z, for the various crystal systems. *Orthorhombic minerals:* X, Y and Z are parallel to a, b and c, but in any order. Thus for one mineral a may be parallel to X, but for other minerals b or c may be parallel to X. *Monoclinic minerals:* the ac-plane contains two optical directions, commonly X and Z, but neither a nor c is parallel to either optical direction. b is parallel to the

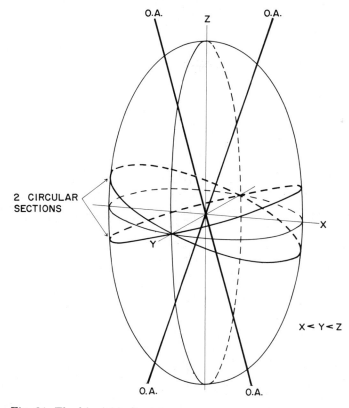

Fig. 21. The biaxial indicatrix.

other optical direction, commonly *Y*. *Triclinic minerals: a, b* and *c* are not parallel to any of the optical directions.

There are always two sections of the biaxial indicatrix that are circular, and these represent planes in the crystal along which light has no preferred vibration direction. Both planes contain and intersect in *Y*, are perpendicular to the *XZ* plane, and are bisected by *X* and *Z* at an angle which depends on the particular mineral. More details of these two sections are given in (4) below.

To illustrate the general properties of biaxial minerals, let us examine the types of section likely to be encountered in laboratory work. The ortho-rhombic mineral olivine is used as an example.

(1) *Mineral section parallel to X and Z of the indicatrix* (Fig. 22). In this case light is travelling parallel to *Y* and is polarised into two rays which vibrate parallel to the fastest and slowest directions *X* and *Z*. The double refraction in this section is the maximum possible and defines the birefringence of the mineral.

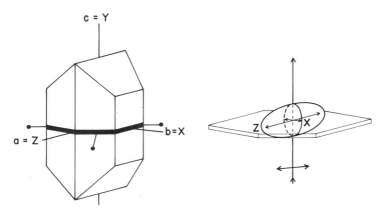

Fig. 22. The biaxial indicatrix and vibration directions in olivine cut parallel to X and Z.

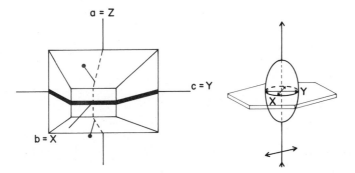

Fig. 23. The biaxial indicatrix and vibration directions in olivine cut parallel to X and Y.

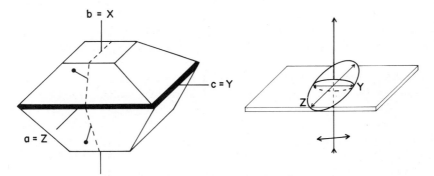

Fig. 24. The biaxial indicatrix and vibration directions in olivine cut parallel to Y and Z.

(2) *Mineral section parallel to X and Y of the indicatrix* (Fig. 23). In this case light travels parallel to *Z* and vibrates parallel to the fastest and intermediate directions *X* and *Y*. Only a partial birefringence is displayed by the section.

(3) *Mineral section parallel to Y and Z of the indicatrix* (Fig. 24). In this case light travels parallel to *X* but vibrates parallel to the intermediate and slowest directions *Y* and *Z*. Only a partial birefringence is displayed by the section.

(4) *Mineral section parallel to either of the two circular sections of the indicatrix* (Fig. 25). The angle between the two circular sections varies from mineral to mineral. We shall continue our examples with olivine in which the two sections commonly make an angle of approximately 90° with each other.

All possible vibration directions of light travelling perpendicular to one of the circular sections lie in that section, and represent the same refractive index. There are no preferred vibration directions. The vibration direction of light from the polariser is therefore unaffected on passing through the mineral and is totally absorbed by the analyser. The section appears black (or in practice often very dark grey), and remains so on rotation of the stage. The partial birefringence of the section is zero.

The direction perpendicular to the circular section is called the optic axis, and there are two such directions for all biaxial minerals (hence biaxial). The angle between the two optic axes is called the 2*V*. When this angle is acute about *Z* the mineral is *positive*, when it is acute about *X* the mineral is *negative*. The optic axes always lie in the *XZ*-plane which is therefore called the *optic axial plane* (O.A.P.). If the mineral is positive, *Z* is called the *acute bisectrix*, and *X* the *obtuse bisectrix*. If the mineral is negative, *X* is the acute and *Z* the obtuse bisectrix.

(5) *Mineral section not parallel to the circular sections and not containing the directions X, Y or Z*. The vibration directions in such a general section of the biaxial indicatrix can be determined by the Biot-Fresnel law which states

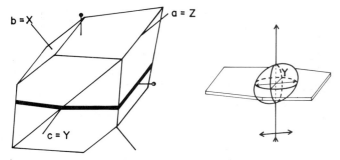

Fig. 25. The biaxial indicatrix and single vibration direction in olivine cut parallel to a circular section.

(approximately) that "the vibration directions bisect the angles between the two planes which are normal to the section and contain one optic axis each". We shall find that this law is important in understanding biaxial interference figures, but it need not concern us during routine laboratory work, since we usually look for sections which contain, or approximately contain, X, Y, or Z.

Interference figures

Normally, we view a mineral section with so-called orthoscopic illumination, or a bundle of approximately parallel light rays. Microscopes are provided with a condenser system (Chapter 2) which when fully raised produces strongly convergent light (called *conoscopic illumination*). The more expensive microscopes are fitted with an extra condenser which can be swung in or out of position whenever necessary. Interference figures are produced by convergent light between crossed-polars, and may be focussed using high-power objectives. To view the figures it is necessary to either use the accessory Bertrand lens, or to remove the eyepiece and insert a pin-hole stop in the top of the microscope tube.

Interference figures are used to determine whether a mineral is uniaxial or biaxial positive or negative; details of their appearance and use are explained in Chapter 4. In this chapter we shall examine the uniaxial cross and the biaxial acute bisectrix figures, in order to understand in principle how they are produced.

Uniaxial cross interference figure

Fig. 26 illustrates a thin section of quartz cut at right angles to its c-axis and placed in convergent polarised light. The rays of light A, B and C pass through the section at oblique angles and are polarised into two rays. The ordinary rays must vibrate in the plane of the section (the circular section of the indicatrix), and since the extraordinary rays vibrate at right angles to both the ω ray and the travel direction of light, they must lie in vertical planes which contain the c-axis. The vibration directions of the ϵ' rays dip in towards the c-axis. All other rays of light passing obliquely through the section are similarly polarised into two rays, and in every case, the ϵ' ray must vibrate in a vertical plane containing the c-axis. Therefore in plan view (Fig. 26c), the vibration directions of the ϵ' rays form a radiating pattern at right angles to which there is a concentric pattern of ω-ray vibration directions. Where these vibration directions are parallel to those of the polariser and analyser (E—W and N—S), light is extinguished completely, and a black cross (called an *isogyre*) is formed (Fig. 26d). The black cross will not change in position when the stage is rotated. Outwards from the centre of the figure, the dip of the ϵ' rays increases, and with it the birefringence. Concentric

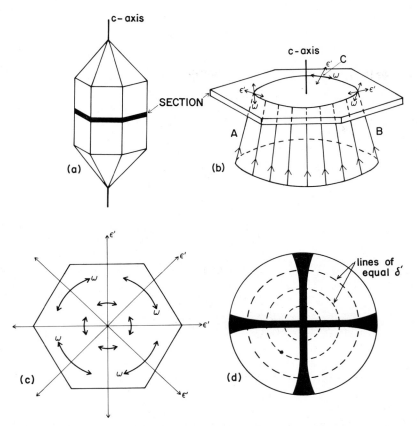

Fig. 26. Illustrating how the uniaxial cross interference figure is produced. See text for explanation.

rings of interference colour are therefore formed, but are only clearly seen with minerals of high birefringence or in thick sections.

Biaxial acute bisectrix interference figure

Fig. 27 illustrates a section of a positive biaxial mineral with a small $2V$ cut perpendicular to Z (the acute bisectrix in positive minerals). The ray of light A passes through the section obliquely and is polarised into two rays, the vibration directions of which can be determined by the Biot-Fresnel law. On the plan view (Fig. 27b), we can draw the traces of the two planes each of which contains an optic axis and the normal to the effective section (the effective section contains the two vibration directions and its normal, of course, is parallel to ray A). The two vibration directions bisect the angles between these two planes, and in this case are parallel to the polariser and

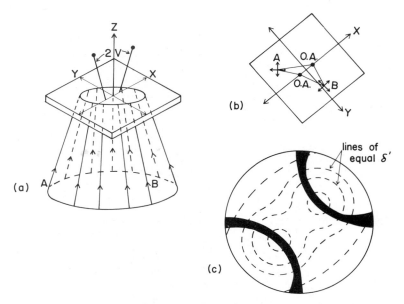

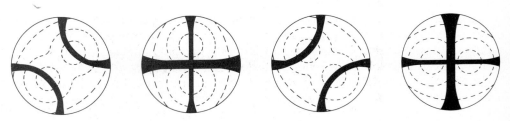

Fig. 27. Illustrating how the biaxial acute bisectrix interference figure is produced. See text for explanation.

analyser vibration directions, and light is extinguished between crossed-polars. The ray of light *B*, on the other hand, is polarised into two rays vibrating at an angle to the polariser and analyser directions, and light is not extinguished. If we determine the vibration directions of all other convergent rays of light, we find that there are two curved bands of extinction (isogyres) passing through the optic axes (Fig. 27c). Birefringence increases outwards from the optic axes, and if the birefringence or thickness of section are sufficiently high for interference colour rings to be present, they form a figure-of-eight pattern (Fig. 27c).

Unlike the uniaxial cross interference figure, the pattern of isogyres

Fig. 28. Changing pattern of the biaxial acute bisectrix interference figure when stage is rotated.

changes as the stage is rotated (Fig. 28). The isogyres join to form a cross when the optical directions X and Y are parallel to the vibration directions of the polariser and analyser, and open up again in the other quadrants on further rotation. The positions of the optic axes are always marked by the figure-of-eight pattern of interference colour rings (if these are present).

CHAPTER 4

Laboratory Techniques

SAMPLE PREPARATION

It is customary to examine rock and mineral samples under the micro-scope either as very thin sections (approximately 0.03 mm thick) or as very fine grains. The main reason for this is that many minerals are not trans-parent except as very small grains, and we must therefore standardise our routine methods of study to deal with them; such a standardisation allows us to examine almost the entire range of grain sizes in rocks, since we can always reduce the size of a large grain by crushing or sectioning. In addition, the important property of birefringence is most easily measured by observing the interference colours in very thin sections of known thickness (using the Michel-Lévy colour chart — Plate 1).

Thin sections and grain mounts both have their particular advantages and disadvantages. Thin sections enable minerals to be identified, and allow grain sizes and textural relationships in rocks to be examined; the section is perma-nent and easily stored. On the other hand, thin sections are relatively expen-sive and time-consuming to make. A crushed-grain mount is the quickest and most economical means of identifying an unknown mineral, and compared to a thin section, allows a much more accurate R.I. determination. On the other hand, no grain-size or textural information is gained, and the mounts are not normally permanent or easy to store. Loose-grain mounts may be the best way to examine sediments that are unconsolidated or easily disaggregat-ed, since they enable the sedimentologist to determine grain size and shape as well as to identify the minerals.

The following is a brief summary of how thin sections and loose- or crushed-grain mounts are made.

Thin sections

(1) A small chip of rock or mineral is sampled, or a thin slice (a few millimetres thick) is cut from the specimen using a diamond saw.

(2) The chip or slice is ground flat and smooth on one surface using progressively finer abrasive, e.g. carborundum powder, starting with 80 grade and finishing with 600 grade.

(3) The smooth surface is cemented to a glass slide (usually 3″ × 1″, but

2″ × 1″ for universal stage work) using canada balsam or Lakeside 70 cement (other cements are occasionally used).

(4) The other surface is then ground down until the rock section is 0.03 mm thick. Progressively finer abrasives are used as the section becomes thinner, and finishing is done with 600-grade carborundum powder. The thickness is gauged by observing the interference colours of common minerals such as quartz or feldspar (a later section of this chapter explains how to judge the thickness).

(5) After cleaning excess cement from the section, it is covered with a glass cover-slip cemented on with canada balsam.

The details of the above method depend on the equipment available. Some sophisticated machines enable sections to be made almost automatically and very quickly. On the other hand, successful sections can be made by hand, the grinding being done slowly on glass plates; indeed, sections of delicate specimens may have to be made in this way.

Some poorly consolidated rocks may require impregnating with a cement before a section can be made. Heating the specimen in canada balsam is sometime sufficient, but stronger cements are available (Reed and Mergner, 1953; Taylor, 1960; Wright, 1964).

Canada balsam is usually thinned with xylol to ease spreading. The mounting of the section is done on a hot-plate which serves to drive off the xylol, and harden the cement.

Loose- and crushed-grain mounts

If a sediment is easily disaggregated, it is customary to spread the grains on a glass slide (often after the grains have been mechanically sorted according to grain size and density), and to immerse them in canada balsam, or less permanently in some liquid. The mount is covered with a glass cover-slip. Sand or finer grain sizes are suitable as loose-grain mounts for direct examination with the microscope.

Crushed-grain mounts are prepared by either scraping mineral fragments from a crystal in a rock or mineral hand-specimen onto a glass slide, or by crushing a small mineral grain on a glass slide using a pocket knife. A suitable immersion liquid is introduced by capillary action from the edge of the cover-slip which is placed over the fragments (see Fig. 30 and the accompanying text for details).

REFRACTIVE INDEX DETERMINATION

1. In thin sections

Relief

A standard thin section is highly irregular in detail (see enlarged part of

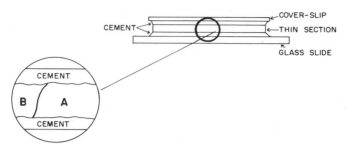

Fig. 29. Thin-section surface irregularities which produce the effect of relief.

Fig. 29). Boundaries between adjacent grains are not generally vertical and may be stepped or curved. The top and bottom surfaces of the section are not polished but pitted and grooved to some degree.

If there is a difference in R.I. between the mineral and the cement, these irregularities concentrate or scatter light by reflection and refraction. The effect is to give an impression of three-dimensional relief. If the difference in R.I. is small, the irregularities will be barely visible and the mineral has *low* relief. A large difference results in *high* relief. Relief should be observed in plane-polarised light with a low- or medium-power objective, and with the diaphragm below the stage nearly closed. Relief may be improved by lowering the condenser system.

The description of relief is somewhat subjective and the apparent relief depends on both the thickness of the section and the degree of irregularity of its surfaces. Nevertheless, we can usefully make estimates of R.I. using relief. Table I presumes canada balsam (R.I. = 1.537) or Lakeside 70 (R.I. = 1.54) to be the cement. This is normally the case, but it is as well for the student to enquire in the laboratory what cement has been used.

The student must practise estimating R.I. from relief, using at first a group of reference thin sections.

TABLE I

Relative relief in thin sections

R.I.	Relief	Mineral example
>1.90	extreme	rutile
1.78—1.90	very high	almandine garnet
1.68—1.78	high	epidote
1.57—1.68	moderate	actinolite
1.49—1.57	low	quartz
<1.49	moderate (less than C.B.)	fluorite

If a mineral has a high or very high relief, it is safe* to say its R.I. is higher than that of canada balsam. However, the R.I. of a mineral with low or moderate relief may be either higher or lower than that of canada balsam, and it is necessary to make a Becke-line test.

Becke-line test

Under the same light conditions as used for observing relief, low- or moderate-relief minerals concentrate light as a thin bright white line along their margins. This line, known as the Becke line, will move inwards or outwards if the mineral grain is brought slowly into or out of focus. The following rule regarding the direction of movement of the Becke line enables us to compare the R.I. of a mineral and canada balsam (e.g. at the edge of the thin section), or the R.I. of two adjacent minerals: the Becke line moves into the substance of higher R.I. when the distance between objective and section is increased. Becke lines cannot be successfully observed with very high-relief minerals.

Twinkling

Some minerals with a high birefringence have one R.I. close to that of canada balsam and the other one considerably different. Such minerals change markedly in relief on rotation of the stage in plane-polarised light, a phenomenon known as twinkling. Calcite is a good example of a mineral that twinkles.

2. In loose- or crushed-grain mounts

A more accurate means of determining R.I. is by immersing loose or crushed grains of a mineral in a series of liquids of known R.I. The relief of the mineral will disappear when its R.I. matches that of the liquid.

A small mineral grain should be crushed to a powder (some experience is necessary since too fine a powder cannot be easily observed whereas it is difficult to immerse too coarse a powder in the liquid). The powder is placed on a glass slide (a minute amount suffices), and is covered with a small glass cover-slip (a round 18 mm diameter size is ideal). Drops of liquid are introduced by capillary action from the edge of the cover-slip until the powder is totally immersed (Fig. 30). By observing relief and Becke lines, the R.I. of the mineral is compared with that of the liquid. This procedure is followed with several liquids until one is found with the same R.I. as the mineral.

Suitable immersion liquids can be bought commercially, but a cheap and

---

* The lowest R.I. of a common rock-forming mineral is 1.43 (fluorite). There are, however, various rare minerals (not described in this book) with much lower R.I. A well-known example is cryolite with an R.I. of 1.34, approximately the same as that of water.

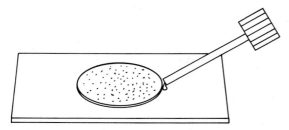

Fig. 30. Immersion liquid introduced at the edge of a cover slip which is placed over mineral fragments.

effective series with a range 1.466—1.778 (with intervals of approximately 0.01) can be made by mixing the following liquids:

1.466—1.633   petroleum oil (Nujol) + alpha-chloronaphthalene
1.633—1.739   alpha-chloronaphthalene + methylene iodide
1.739—1.778   methylene iodide + methylene iodide saturated with sulphur

Alpha-bromonaphthalene (R.I. = 1.655) may be used instead of alpha-chloronaphthalene.

A variety of volatile liquids, including some of the alcohols and petroleum distillates, may be used for a lower range of R.I. (Harrington and Buerger, 1931; Larsen and Berman, 1934; Weaver and McVay, 1960). Liquids with R.I. higher than 1.778 are generally toxic, corrosive, and difficult to handle. Such liquids are readily available commercially (e.g. Cargille Labs. Inc., Cedar Grove, N.J., U.S.A.). One high-index series can be made as follows:

1.655—1.814   alpha-bromonaphthalene + 10% S in $AsBr_3$
1.814—2.00   10% S in $AsBr_3$ + 20% S and 20% AsS_2 in 60% $AsBr_3$

Seldom are higher-index liquids required, but they may be prepared from various low-temperature melting compounds (Larsen and Berman, 1934; Meyrowitz, 1955).

The lower-R.I. liquid mixtures can be measured and periodically checked using a standard refractometer. The Abbe type will measure liquids in the range 1.3—1.7, and the Leicz-Jelley type liquids in the range 1.3—1.9.

The simple immersion method described here is normally adequate for routine mineral identification. If the mineral has a moderate to high birefringence, it may be necessary to measure ω, ϵ, α, β or γ separately; a suitable procedure is described in the next chapter under loose- and crushed-grain mounts. If an extremely accurate R.I. measurement is required, more sophisticated techniques using monochromatic light and thermostatic controls must be employed (Emmons, 1943; Wahlstrom, 1969).

DETERMINATION OF BIREFRINGENCE

1. In thin sections

The birefringence or partial birefringence of a mineral section is determined by observing interference colours in crossed-polarised light with the
diaphragm below the stage open, and using a low- or moderate-power objective. Fig. 31 diagrammatically represents the Michel-Lévy colour chart (Plate
1). Values of birefringence are represented by lines radiating from the lower
left hand corner of the diagram. Thickness of the mineral section is represented by horizontal lines, and the interference colours are vertical lines.
Basically the technique involves determining: (1) the thickness of the section, and (2) the interference colour. For example, a mineral section with
interference colour A and thickness B (Fig. 31) has a birefringence or partial
birefringence C. To identify an unknown mineral, we are normally concerned with determining the true birefringence, that is the difference between ω and ϵ, or α and γ.

Choosing a grain for determination of true birefringence

The several grains of a mineral in a thin section will normally be cut
parallel to different crystallographic directions, and hence intersect the indicatrix in different orientations. The interference colour will vary from black
(or very dark grey) for those cut parallel to a circular section of the indicatrix, to a maximum for those cut parallel to ω and ϵ, or X and Z. Thus for a
mineral with birefringence E in a section thickness D we may see a number
of grains with interference colours ranging from 1 to 7 (Fig. 32). To measure
the true birefringence we must therefore search for the grain with the highest
interference colour (e.g. grain 7 in Fig. 32). The search must be thorough. A
problem arises if the mineral has a preferred orientation (as is often the case
in metamorphic rocks), when suitably oriented sections may not be present.

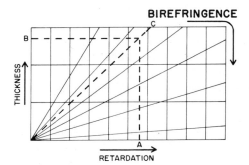

Fig. 31. Diagram illustrating how the Michel-Lévy chart (Plate 1) is used.

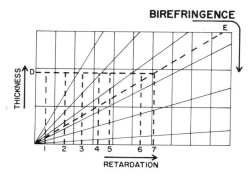

Fig. 32. Illustrating a number of possible interference colours produced by a mineral with birefringence E in a section of thickness D but with a number of orientations.

Identifying the interference colour

The interference colours are normally divided into *orders*, 1st, 2nd, 3rd, 4th, etc., corresponding to the amount by which the two rays of light are out of phase. A wavelength of 560 nm is chosen as a convenient standard. It will be noticed (Plate 1) that some colours repeat themselves on the interference colour chart, so that there is a 2nd- and 3rd-order red, for example. These repeated colours become paler with increased order, but the student must beware of determining the order from the colour intensity, or by direct comparison with the chart.*

The simplest method to determine the order of colour is to look at the edge of mineral grains. Grains are normally wedge-shaped at their margins, and as the thickness reduces so will the interference colour. For example, if a section with birefringence F reduces in thickness at its edge from G to H, the interference colour will change from I to J as a series of colour rings (Fig. 33). It is nearly always possible to observe the unique and distinctive 1st-order white or grey at the edge of a grain that displays colour rings (using a moderate- or high-power objective). By counting the colour rings, the order of colour in the main part of the mineral section may be determined. Only

---

* *Anomalous interference colours* not represented on the Michel-Lévy chart are observed occasionally, especially within the 1st-order range of colours (e.g. blues and purples instead of 1st-order grey and white). There are two main causes: (1) the body colour of the mineral may affect the interference colour, (2) the optical properties of the mineral may differ for varying wavelengths of light. Thus the positions of X, Y and Z may differ for the red and blue parts of the spectrum. This phenomenon is known as *dispersion*, and may result in anomalous interference colours. Minerals likely to show anomalous colours include chlorite, members of the epidote family, vesuvianite, and melilite. The birefringence represented by an anomalous colour may usually be ascertained with the sensitive-tint accessory plate (q.v.) which will cause compensation or retardation by 560 nm producing a normal and recognisable interference colour.

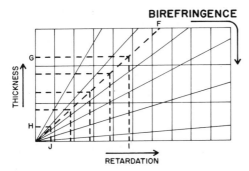

Fig. 33. Illustrating the change of interference colours with change of thickness in a section with birefringence F.

when the mineral has an extreme birefringence, and hence a very large number of colour rings, will this technique not work.

Determination of thin-section thickness

Standard modern thin sections are close to 0.03 mm in thickness. It is easy for the student to confirm this provided a mineral of known birefringence has already been identified in the section. Quartz is ideal for the purpose since it is common, easily identified, and of constant birefringence. Feldspar and other minerals may be used but with less accuracy.

Quartz has a birefringence of 0.009. By reference to Plate 1, it can be seen that in a section of 0.03 mm thickness there should be a few grains with 1st-order pale-yellow interference colours. If there are none, the section is thinner than normal; if there are grains with orange interference colours, it is thicker than normal. With some experience it is possible to determine the thickness within 0.01 mm (note, however, that a thin section may vary a little in thickness).

Summary of method of birefringence determination

To measure the true birefringence of a mineral, a thorough search must be made for a grain with the highest interference colour. The birefringence may be read from the Michel-Lévy chart once the thickness of the thin section has been checked against quartz or some other mineral whose birefringence is already known.

2. In loose- or crushed-grain mounts

Birefringence cannot be measured so accurately in loose-grain mounts because of the presence of grains of differing (and often unknown) thickness. An additional problem is the preferred orientation taken up by euhedral or well-cleaved grains which may result in the maximum birefringence section not being observed. The method of birefringence measurement is the same as for thin sections, provided these problems can be overcome.

FAST AND SLOW DIRECTIONS AND THE USE OF ACCESSORY PLATES

All petrographic microscopes are provided with one or more accessory plates. The most useful of these is the sensitive-tint plate, also known as a gypsum or 1-wavelength plate. This consists of a section of gypsum or quartz cut with a thickness and orientation such that it produces an interference colour of 1st-order pink (approximately 1 wavelength (1 λ) out of phase for the standard 560 nm). This colour may be observed simply by inserting the plate in the accessory slot in the tube of the microscope, using crossed-polarised light. When inserted, the plate has two vibration directions in the 45° positions and these are marked on the plate as fast and/or slow (or X and Z or α and γ respectively). Other plates commonly provided are the mica plate ($\frac{1}{4}$ λ) producing a 1st-order grey interference colour, and the quartz wedge which has a variable thickness producing a gradational series of inter-ference colours.

The accessory plates are used for: (1) increasing the retardation produced by a mineral section (in other words, it is equivalent to increasing the thick-ness of a section), or (2) compensating for the wavelength difference pro-duced by a mineral. This is achieved as follows: first put a mineral section in extinction (Fig. 34a), thus placing its two vibration directions E—W and N—S; then rotate 45° (Fig. 34b) noting whether clockwise or anticlockwise, and determine the interference colour produced by the mineral; then insert the accessory plate. If further retardation occurs (i.e. the order of interfer-ence colour increases), then the slow direction of the plate is parallel to the slow direction of the mineral (i.e. in Fig. 34b, the vibration directions A and B are slow and fast respectively). If compensation occurs (i.e. the order of interference colour decreases), then the slow direction of the plate is parallel

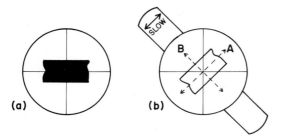

(a)   (b)

Fig. 34. Use of an accessory plate to locate fast and slow directions. See text for explana-tion.

to the fast direction of the mineral (i.e. in Fig. 34b, the vibration directions *A* and *B* are fast and slow respectively).

The amount of retardation or compensation (also known as "addition" and "subtraction") will be by approximately 560 nm for a 1-λ plate or 140 nm for the mica plate. Sometimes compensation of a low interference colour results in an apparent retardation. For example 1st-order white (200 nm) is changed by compensation to 1st-order yellow (360 nm) using a 1-λ plate (560 nm). One must view this as the compensation of white to zero combined with a retardation of 360 nm which is the excess retardation of the 1-λ plate. Care must be taken to correctly identify the interference colour before and after insertion of the plate. This is best done by observing the colour rings on the thin edge of grains.

Determination of the fast and slow directions with respect to distinctive crystal faces or good cleavages is useful in identifying minerals. We shall learn how to specify optical directions with more precision in the section on orientation diagrams.

INTERFERENCE FIGURES, DETERMINATION OF OPTIC SIGN, AND THE MEASUREMENT OF 2V

Interference figures are obtained as follows: focus on a grain with a high-power objective (× 40 or greater); make sure the grain is central in the field of view, and that the microscope is centred perfectly (see Chapter 2); raise the sub-stage condenser, insert the accessory condenser (if the microscope has one) and open the sub-stage diaphragm; use crossed-polarised light; insert the Bertrand lens, or remove the eyepiece and insert a pin-hole stop; if there is a diaphragm in the microscope tube, partially close it. The stage should be rotated when observing interference figures so that the moving pattern of isogyres can be noted.

Interference figures are used principally to determine whether the mineral is uniaxial or biaxial, and with the aid of accessory plates, to determine whether the mineral is +ve or —ve. They are also used to determine the 2*V* of biaxial minerals. *All* these properties can *always* be determined from grains cut parallel to the circular section(s) of the uniaxial or biaxial indicatrix. Therefore, always look for a grain with the lowest interference colour (black or dark grey) for that mineral — such a grain will produce either a uniaxial cross or a biaxial optic-axis figure (see below). Note that a cubic mineral will be black in crossed-polarised light, but will not produce an interference figure.

The interference figures from other grains are of some use, especially in constructing orientation diagrams (q.v.).

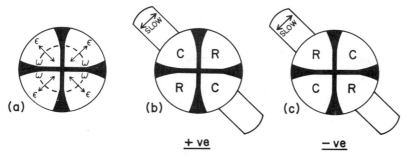

Fig. 35. Patterns of retardation and compensation in uniaxial cross interference figures using accessory plates. R = retardation, C = compensation.

Use of accessory plates with interference figures

To determine the optic sign and the position of the optical directions in minerals it is necessary to examine interference figures with the aid of accessory plates.

In uniaxial cross interference figures, the ϵ-vibration directions are radial and the ω directions concentric (Fig. 35a), and insertion of the plate results in retardation or compensation in alternate quadrants. It can be recalled here that the mineral is positive if ϵ is slow and negative if ϵ is fast. Thus the pattern of retardation and compensation distinguishes positive from negative minerals (Fig. 35).

In biaxial bisectrix interference figures, the pattern of retardation and compensation in relation to the isogyres enables us to determine the positions of X, Y, and Z. For example, Fig. 36 shows the patterns of retardation and compensation produced in figures with X vertical (note that Y is always perpendicular to the optic axial plane).

The patterns of retardation and compensation in all optic-axis and some bisectrix figures enable us to determine the optic sign of the mineral (see below).

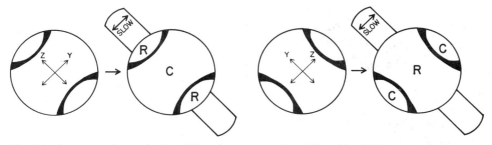

Fig. 36. Patterns of retardation (R) and compensation (C) in biaxial bisectrix interference figures from a YZ section using accessory plates.

To determine the patterns of retardation and compensation, we can use either the sensitive-tint plate or the quartz wedge, as follows:

Use of sensitive-tint plate

With standard thickness sections of minerals with a low or moderate bire-fringence, 1st-order white is the dominant or conspicuous interference col-our between the isogyres of an interference figure. Even when several colour rings are present in interference figures from high-birefringence minerals, a small area of 1st-order white can be seen immediately adjacent to the posi-tion of an optic axis. In most cases therefore, the pattern of compensation and retardation is easily determined by inserting the sensitive-tint plate and noting the change in interference colour of 1st-order white to either 1st-order yellow (compensation) or 2nd-order blue (retardation). Alternatively, the change in order of the colour rings, if present, may be observed.

Use of the quartz wedge

If a large number of colour rings are present it may be necessary to use a quartz wedge. As the wedge is inserted, there is a gradual change in the interference colour so that the colour rings appear to move in towards an optic axis (retardation) or to move away from an optic axis (compensation). As the wedge is removed, the rings move, of course, in the opposite direc-tion.

Interpretation of interference figures

In the following diagrams, b and y refer to the 2nd-order blue and 1st-order yellow interference colours produced by retardation and compensation (from 1st-order white) respectively, on insertion of a sensitive-tint plate. The arrows refer to the movement of colour rings when the quartz wedge is inserted. All diagrams assume the accessory plates have their slow directions NE—SW (or fast NW—SE). If plates are used with a slow direction NW—SE, all the colour and arrow schemes should be reversed.

The isogyres appear red when the sensitive-tint plate is used.

Uniaxial cross interference figures (Fig. 37)

Interpretations possible: Mineral is uniaxial or biaxial with $2V$ approxi-mately $0°$. The optic axis or axes are approximately vertical in the section. +ve or —ve character may be determined.

Notes: Off-centred figures can be used provided the centre of the cross can always be seen (Fig. 38). In these cases the optic axis may be as much as $20–30°$ from vertical in the section. Off-centred figures in which straight isogyres move across the field of view but with the cross outside the field of view should not be used since they cannot be easily distinguished from some biaxial figures.

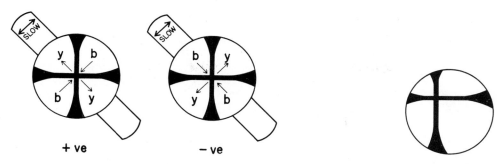

+ ve   − ve

Fig. 37. Distinction of uniaxial +ve from uniaxial —ve figures using a sensitive-tint plate or quartz wedge. b = blue; y = yellow; arrows indicate movement of colour rings when wedge inserted.

Fig. 38. Off-centred uniaxial cross interference figure.

Biaxial optic-axis figures — An isogyre remains centred in the field of view during rotation (Fig. 39)

Interpretations possible: Mineral is biaxial, and an optic axis is vertical in the section. The degree of curvature of the isogyre enables $2V$ to be *estimated* (a correction for R.I. of the mineral and the numerical aperture of the objective is strictly necessary). If $2V$ is 90°, the isogyre remains straight during rotation. The curvature becomes greater with decreasing $2V$ (Fig. 39), and when the $2V$ is less than approximately 20°, an additional isogyre is seen towards the edge of the field of view. If the $2V$ is 0° or close to 0°, the figure is similar to or indistinguishable from the uniaxial cross. +ve or —ve character can always be determined (Fig. 40). Remember that if the $2V$ is 90°, the mineral is neither +ve nor —ve.

The figure can be used to determine the positions of Y and an optic axis for orientation diagrams.

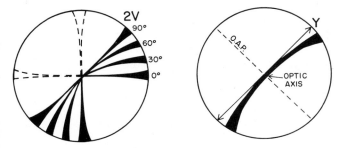

Fig. 39. Measurement of $2V$ and location of Y and O.A.P. using the biaxial optic-axis figure.

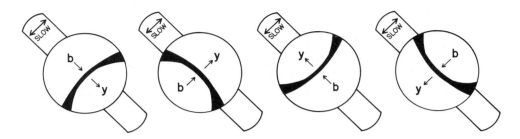

All +ve. Schemes reversed for −ve.

Fig. 40. Optic-axis figure of a biaxial +ve mineral using a sensitive-tint plate or quartz wedge. b = blue; y = yellow; arrows indicate movement of colour rings when wedge inserted.

Acute bisectrix figures (small to moderate 2V) — Isogyres remain in field of view when stage is rotated (Fig. 41)

Interpretations possible: Mineral is biaxial. The acute bisectrix is near vertical in the section. +ve or −ve character may be determined. $2V$ may be *estimated* from the maximum separation of the isogyres in the 45° position. The isogyres barely separate if the $2V$ is very small (less than 10°), whereas if the $2V$ is of the order of 50° the isogyres move to the edge of the field of

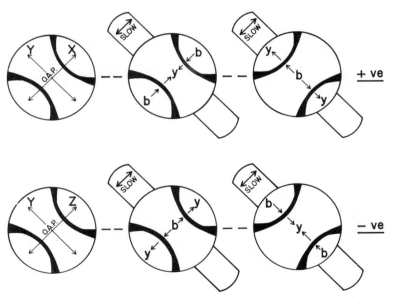

Fig. 41. Distinction of biaxial +ve from biaxial −ve acute bisectrix interference figures using a sensitive-tint plate or quartz wedge. b = blue; y = yellow; arrows indicate movement of colour rings when wedge inserted.

view. The estimations possible are useful in most routine work, but to determine $2V$ accurately it is necessary to make corrections for the R.I. of the mineral* and the optics of the microscope** (Tobi, 1956), or to use universal stage methods (see later).

The figure can be used to determine the position of X, Y and Z for orientation diagrams.

Note: Off-centred figures may be used provided the isogyres clearly cross, and always stay in the field of view.

Acute and obtuse bisectrix figures (moderate to large 2V) — Isogyres move slowly out of field of view when stage is rotated (Fig. 42)

Interpretations possible: Mineral is biaxial. The acute or obtuse bisectrix is near vertical, but they cannot be distinguished. Therefore, the +ve or —ve character of the mineral cannot be determined. The $2V$ is moderately large

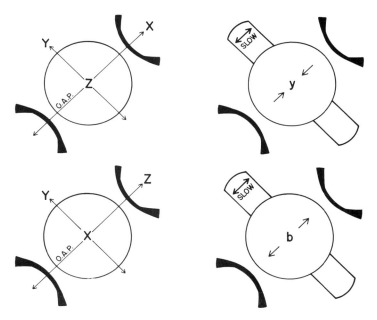

Fig. 42. Use of sensitive-tint plate or quartz wedge in bisectrix figures when isogyres move slowly out of field of view. b = blue; y = yellow; arrows indicate movement of colour rings when wedge inserted.

---

* This is because the angle between the rays of light travelling parallel to the two optic axes is increased by refraction from the section into air. The actual angle observed in bisectrix figures is termed $2E$, and is related to $2V$ by the formula $\beta \sin V = \sin E$.
** The apparent separation of the isogyres is governed by the numerical aperture (N.A.) of the objective.

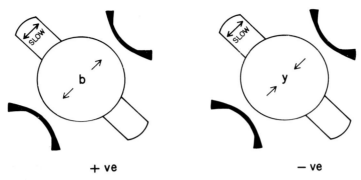

Fig. 43. Flash figures from +ve and —ve minerals using a sensitive-tint plate or quartz wedge. b = blue; y = yellow; arrows indicate movement of colour rings when wedge inserted.

or large. Can be used to determine the positions of X, Y and Z for orientation diagrams.

Notes: Off-centred figures may be used provided the isogyres clearly cross, and move slowly out of the field of view. It is necessary to obtain an optic-axis figure to determine $2V$ and the +ve or —ve character.

Obtuse bisectrix figure (small to moderate 2V) and uniaxial flash figures — Isogyres move quickly out of field of view when stage is rotated (Fig. 43)

These figures are called flash figures because the isogyres move out of the field of view very quickly (within approximately 30° rotation of the stage).

Interpretations possible: May be used to determine whether mineral is +ve or —ve, although it is advisable to check the determination with an optic-axis figure. Optic axes lie in or close to the plane of the section.

Notes: The figure does not distinguish biaxial and uniaxial minerals. It is necessary to obtain an optic axis or an acute bisectrix figure to determine $2V$.

Figures produced when X and Z lie parallel to the section

If the $2V$ is small, a flash figure similar to that described above is seen. If the $2V$ is large, a flash figure is produced in which the direction of separation of the isogyres cannot be seen clearly. Such centred figures may be useful in confirming the presence of an XZ section.

CRYSTAL SHAPE AND CLEAVAGE

Great care must be exercised in determining the shape of crystals from thin-section studies. For example, the two-dimensional appearance of an oblique section of a square prism is a rhomb (Fig. 44).

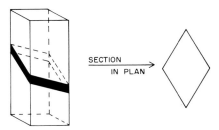

Fig. 44. Rhomb-shaped section from a square prism.

Neither should it be forgotten that in many rocks the crystal boundaries are not crystal faces but mutual growth boundaries with adjacent grains. This is particularly true in metamorphic rocks where the shape of grains in a mosaic may bear no relationship to crystallography.

Nevertheless, euhedral crystal shapes and habits can be judged in a *approximate* way if sufficient care is exercised, and they may be useful in identification of the mineral.

Cleavage is commonly observed as a set of cracks or bands in a mineral section. Cleavage vertical to the section appears as fine cracks, but cleavage oblique to the section will appear as dark broad bands. Cleavages more than $30°$ from vertical to the section are not normally seen at all. Information on the number of cleavages and their relationship to optical directions and crystal shapes is of considerable importance in the identification of minerals. It is useful to remember that cleavages are nearly always parallel to simple crystal faces e.g. (100), (110), etc., and that the presence or lack of cleavage in certain directions may reflect crystal symmetry (see Chapter 1).

To make the best use of crystal shape and cleavage information, orientation diagrams should be constructed, as described in the next section.

ORIENTATION DIAGRAMS AND EXTINCTION ANGLES

Extinction angles

The angle between a vibration direction and a cleavage or prominent crystal face is called the extinction angle. If the angle is zero, the mineral is said to have straight extinction. If the angle is not zero, the extinction is inclined; sometimes extinction is symmetrical to the cleavages or crystal faces. The angle is determined by putting the grain in extinction and then rotating the stage until the cleavage or crystal face is parallel to the crosshairs (polariser and analyser vibration directions), and noting the angle of rotation. This is a most useful property provided it is properly related to precise optical directions. This can be achieved by constructing orientation diagrams.

Orientation diagrams

Throughout the systematic description of minerals in this (and other) books, orientation diagrams are employed to illustrate optical properties. It is essential that the student learns to use them, and that he is capable of constructing them for himself. There are two basic approaches to their construction.

1. Using characteristic optical sections

A search is made for the sections that show the highest and lowest interference colours for the mineral. These are sought in any case for the determination of birefringence and the nature of the indicatrix.

(A) *Lowest interference colour section*. Such a section is cut parallel to one of the circular sections of the biaxial indicatrix or perpendicular to the c-axis of a uniaxial mineral. Proceed as follows:

(1) Obtain an interference figure. Determine whether uniaxial +ve or —ve or biaxial +ve or —ve (this will always be possible if the correct section is chosen). If biaxial, estimate $2V$.

(2) If uniaxial, remove Bertrand lens, and view in plane-polarised light. Draw the section, noting any crystal shape (if euhedral, the section should be a basal one reflecting the hexagonal, trigonal or tetragonal symmetry) and/or cleavages present.

(3) If biaxial, rotate the interference figure into the 45° position. In such a position the orientations of the O.A.P. and Y can be specified (Fig. 45). Without rotating the stage, remove the Bertrand lens and view the section in plane-polarised light. Draw the section noting any crystal shape and position of cleavages if present. Superimpose on the drawing the position of Y and the O.A.P. as determined from the interference figure.

(B) *Highest interference colour section*. Such a section is parallel to X and Z of a biaxial mineral or parallel to the c-axis of a uniaxial mineral. Proceed as follows:

(1) Put the grain into extinction, remove the analyser and draw the section, noting any crystal shape and/or cleavages. The two optical directions can be marked as being parallel to the cross-hairs.

(2) If the mineral is uniaxial we should already know from section A

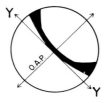

Fig. 45. Use of biaxial optic-axis figure for specifying Y and position of O.A.P.

(above) whether the mineral is +ve or —ve. If uniaxial +ve then the direction parallel to the c-axis (ϵ) is slow, if —ve it is fast. Replace the analyser, rotate the stage 45° and using an accessory plate determine the position of the c-axis by noting the fast and slow directions. Uniaxial minerals *always* have straight or symmetrical extinction with respect to crystal shapes and cleavages.

(3) If the mineral is biaxial, replace the analyser, rotate the stage 45° and using accessory plates determine which direction is fast and which is slow. Mark these on the drawing as X and Z respectively.

Euhedral minerals belonging to the orthorhombic system will have straight or symmetrical extinction with respect to crystal shapes and cleavages. There will normally be inclined extinction in triclinic minerals. If X or $Z = b$ of a monoclinic mineral, there will be straight or symmetrical extinction, but since Y usually $= b$, there is usually inclined extinction in XZ sections.

2. Using characteristic crystallographic sections

A search is made for sections which are parallel or perpendicular to prominent crystal faces or cleavages. If the mineral is uniaxial, these sections will usually be the same as the two characteristic optical sections already studied under 1 (above). Since optical and crystallographic directions have more complex interrelationships in biaxial minerals, their characteristic crystallographic sections will not necessarily be the same as those studied under 1 (above).

Therefore for biaxial minerals proceed as follows:

(1) Look for sections that may be judged as parallel to important crystal faces, or that display cleavages perfectly vertical to the section. (N.B. that if a mineral has one good cleavage, there should be sections parallel to it which do not appear to have a cleavage. The presence of such sections should be anticipated, and a careful search made for them. Similarly, minerals with two cleavages should be observed not only in sections displaying the two cleavages, but also in those displaying only one).

(2) Put the mineral into extinction, remove the analyser and draw the grain illustrating the crystal faces (if any) and cleavages (if any), and superimpose the orientation of the cross-hairs (which indicate the positions of two vibration directions).

(3) Replace the analyser, rotate the stage 45°, and using an accessory plate determine the fast and slow directions and note them on the drawing.

(4) Obtain an interference figure.

Sometimes one is fortunate enough to obtain a centred bisectrix figure in which case the position of X, Y and Z can immediately be noted on the drawing of the mineral section. An example of the procedure is given below.

(a) The mineral was drawn in its extinction position (Fig. 46a), and by later rotation and use of an accessory plate, the fast and slow directions identified and noted.

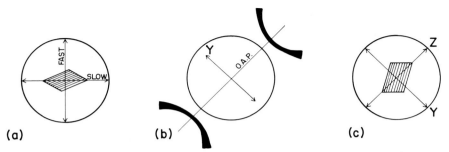

Fig. 46. Example showing how an orientation diagram may be constructed.

(b) A bisectrix interference figure was obtained in which the isogyres rotated out of the NE and SW quadrants. We can specify therefore the positions of the O.A.P. and Y (always perpendicular to the O.A.P.) (Fig. 46b).

(c) Without rotating from this position the analyser and Bertrand lens were removed and the grain observed. (Fig. 46c).

(d) The slow direction in this case lies in the O.A.P. and must therefore be Z. The section also contains Y. X is therefore vertical in the section.

Even if a non-centred figure is obtained, by combining the determination of slow and fast directions and the partial birefringence of the section, sufficient data may be accumulated to allow comparison with the diagrams in the systematic section of this book. If non-centred figures are always obtained on characteristic crystallographic sections, this in itself is significant, and usually implies that the mineral has a triclinic symmetry.

COLOUR AND PLEOCHROISM

To observe colour, the analyser of the microscope must be removed.

The vast majority of minerals are transparent in thin sections or as small grains. Many that are coloured in hand specimen are colourless in thin section. A few minerals are opaque in thin section, and require special methods of study using reflected light.

Whilst the colour, or lack of it, is not by itself diagnostic of a mineral, it is an important property, and usually the first to be observed under the microscope.

Opaque minerals

Ideally, a polished section should be made and the mineral then observed with a special reflected light microscope, such as are used by ore mineralogists (Galopin and Henry, 1972). With an ordinary microscope, a limited amount of information can be obtained by shining a light over the specimen.

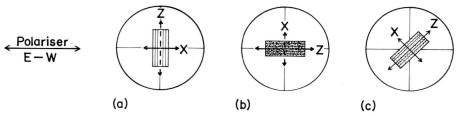

(a) (b) (c)

Fig. 47. Pleochroic biotite in plane-polarised light: $Z > X$.

Determinative Table X should be referred to for the colours of the common opaque minerals.

Transparent minerals

Minerals that are commonly coloured in thin section are listed in Determinative Table IV. Anisotropic minerals frequently display variations in colour or intensity of colour when the stage is rotated in plane-polarised light, a phenomenon known as *pleochroism*. This can be understood if we remember that light vibrates in two particular directions in any one section. Crystals differentially absorb light passing through them depending on the vibration direction of the light.

A good example of a strongly pleochroic mineral is biotite in which the X optical direction absorbs light weakly and the Y and Z directions absorb light strongly. The effect of this is illustrated in Fig. 47. In plane-polarised light in Fig. 47a, light can only vibrate parallel to X and we see a pale-brown colour. When the crystal section is orientated as in Fig. 47b, light only vibrates parallel to Z and we see a dark-brown colour. In intermediate positions (Fig. 47c), light from the polariser is resolved to vibrate in both the X and Z directions, and we see an intermediate shade of brown.

To determine the pleochroic scheme of a mineral proceed as follows:

(1) Draw an orientation diagram, as outlined in the previous section.

(2) In plane-polarised light, rotate the grain so that ω or ϵ, or X, Y or Z are in turn parallel to the polariser vibration direction. Note the colour and its intensity for each optical direction that has been identified.

UNIVERSAL-STAGE METHODS

Application

The universal stage (U-stage) is a device by which thin sections or permanent grain mounts may be rotated about any desired axis. The U-stage is not normally required for routine mineral identification, but its use can assist students in their understanding of optical principles. Some of the more important applications of U-stage methods to advanced work in petrology and mineralogy are as follows.

The positions in space of the uniaxial optic axis and the biaxial optical directions *X*, *Y* and *Z*, can be located precisely and related to such crystallographic features as cleavage and twinning. The exact angle of 2*V* can be measured. This information may be useful for *identifying problematical minerals*, and is essential for *reporting precise mineralogical data*; it may also be useful in making *modal analyses* of rocks when the distinction of two minerals such as quartz and feldspar cannot be made easily for each grain in the specimen. The information can also be applied to the *determination of anorthite content and the twin-laws of plagioclase* (Slemmons, 1962), and to the *distinction of the trigonal carbonates* (Wolf et al., 1967). The U-stage is used to determine the preferred orientation of mineral grains in a rock. Briefly, the method involves determination of the orientation of the optical indicatrix for a large number of mineral grains in a single thin section; the results are plotted on a stereogram to assess the degree of any preferred orientation. The reader is referred to Fairbairn (1949) for details.

The equipment

The U-stage described here (Fig. 48) is that manufactured by Leitz, and has four axes of rotation. U-stages with three or five axes are also available. In fact, three axes suffice for most purposes.

Fig. 48. Right: four-axis universal-stage manufactured by Leitz Wetzlar, Germany. Centre, from top to bottom: circular glass plate; lower hemisphere; upper hemisphere. Left: upper hemisphere with a square mount and slide guide, used for petrofabric work.

The axes of rotation

With the central glass plate of the U-stage parallel to the microscope stage, the four axes can be designated as two "vertical" axes of rotation, A1 and A3, and two "horizontal" axes, A2 and A4.

A1 inner stage ⎫
A3 outer stage ⎬ initially vertical

A2 N—S axis of tilt of inner ring ⎫
A4 main E—W control axis ⎬ initially horizontal

The microscope stage axis is also vertical and is termed A5. A3 is not normally used, and it is best kept permanently clamped in zero position.

The hemispheres

Tilting a mineral section in a path of light soon results in a high degree of refraction and reflection which makes observations impossible. To avoid such reflection it is necessary to use glass hemispheres (Fig. 48) both below and above the mineral section. If there is to be no deviation of the light ray on passing through the mineral and the hemispheres, the R.I. of the glass should equal that of the mineral. Glass hemispheres are manufactured with a variety of R.I., and they should be chosen to match that of the mineral to be studied. Corrections to readings necessary because of a difference in R.I. are easily made, and must always be made if the angle of tilt is greater than 30—40° or the difference in R.I. is greater than 0.10. The study of minerals such as calcite, which have a very high birefringence, may necessitate corrections to be made to U-stage measurements. Errors due to small differences in R.I. of mineral and hemisphere are not normally large enough to cause concern in routine work, and are tolerated.

Corrections. These are simply made using the well-known Federow diagram (Fig. 49). The way in which it is used is shown in the inset diagram. The observed angle is plotted along a radial line to the point where it reaches a concentric line representing the R.I. of the mineral. A vertical line is then drawn up to or down to the concentric line that represents the R.I. of the hemisphere, and the radial line intersecting this point gives the corrected angle.

Objectives

Unusually large working distances between the section and objectives are involved when working with the glass hemispheres. Special objectives must be used. As noted below, it is desirable that these objectives be fitted with individual diaphragms.

Illumination

Sharp extinction positions must be obtained to measure accurately the optical directions of minerals. An intense source of light is required. The

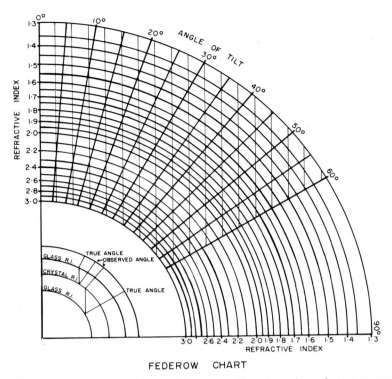

FEDEROW   CHART

Fig. 49. Federow chart for making corrections to universal-stage measurements.

diaphragms below the stage and within the special U-stage objectives are used to produce a good parallel beam of light, and to avoid reflection from the hemispheres.

Setting up a specimen

A small drop of glycerine is placed on the circular glass plate supplied with the U-stage. The lower hemisphere is placed on the drop, thus sealing the contact area from air. The hemisphere and plate are then placed on the U-stage; a spring fitting prevents the hemisphere from falling. The thin section is placed on the glass plate and the upper hemisphere screwed into place on top, both contact surfaces of the thin section being sealed with glycerine. If the thin section has to be kept in a particular orientation (e.g. for petrofabric work), an upper hemisphere with a square edge mount is invaluable. This special mount may be fitted with a slide guide (Fig. 48).

Adjusting the U-stage involves the following steps:

(1) Centre the microscope stage by rotating A5 and adjusting the centring screws of the microscope.

(2) Clamp A5, then rotate A1 and centre, using the centring srews on the U-stage.

(3) Rotate A4 and A2. It is necessary that these rotation axes lie in the plane of the thin section. If they do not, the image will appear to rock backwards and forwards on rotation. Adjustment is made by turning the threaded mount of the lower hemisphere (be careful not to mark the hemispheres with your thumb!).

(4) Finally, focus on some irregularity or pitting on the top of the upper hemisphere. Adjust A5 to a position so that when A4 is rotated the surface irregularities move precisely parallel to the N—S cross-hair. Note the position of A5 and clamp.

Recording measurements

Readings are recorded on tracing paper placed over a stereographic net. A small reference mark should be made on the tracing paper to coincide with the index mark on the A1 axis ring, which is the S pole on the Leitz stage. An introduction to stereographic projection is not given here since most students with a grounding in crystallography and/or structural geology will already be familiar with it. Those who are not, are referred to texts such as Phillips (1971), Fairbairn (1949), or Dana (1932). Data is normally recorded on the lower hemisphere of the stereogram.

Measurement of planar features

To measure the orientation of cleavage cracks and twin-lamellae in a crystal requires the use of only A1 and A4. Therefore A2, A3 and A5 are kept clamped in zero position.

Each feature to be measured should first be rotated parallel to the E—W cross-hair using A1. Rotation about A4 is then made until the cleavage planes or twin boundaries are seen as the sharpest and finest possible lines. With a pleochroic mineral such as biotite, it is sometimes advisable to rotate the polariser of the microscope so as to lighten the colour of the mineral.

In noting readings be careful to note whether A4 was rotated away or towards you (e.g. by an arrow such as 30°↑ or 30°↓).

To plot a reading such as (A1) 41°, (A4) 29°↓, rotate the tracing paper until the index mark reads 41° on the perimeter of the net. The pole to the planar feature is plotted 29° along the N—S diameter of the net from the north side of the perimeter (for lower hemisphere of the stereogram).

Determination of the optical indicatrix of a mineral

Distinction of uniaxial from biaxial minerals

The aim of standard U-stage procedures is to identify the symmetry axes of the optical indicatrix of a mineral. The number of axes differ for uniaxial and biaxial minerals.

Uniaxial minerals. The optic axis and all directions at right angles to the optic axis are symmetry axes of the indicatrix.

Biaxial minerals. The optical directions X, Y and Z are the only symmetry axes of the indicatrix.

Distinction of uniaxial and biaxial minerals can normally be made as follows. If a uniaxial mineral is put into extinction by rotation about A1, in every case a symmetry axis will lie parallel to either A2 or A4, and the grain will remain extinguished when one of either A2 or A4 is rotated. If a biaxial mineral is similarly extinguished by rotation about A1, rotation about both the A2 and A4 axes will result in illumination except in the special cases where X, Y or Z lies precisely parallel to A2 or A4.

The following sections outline the detailed procedure for determining the orientation of the indicatrix of uniaxial and biaxial minerals. Flow-charts have been devised by Flinn (1973) as a practical alternative to the following written instructions.

Determination of the optic axis (c-axis) of a uniaxial mineral

The optic axis (c-axis) is the only direction that can be determined by examining the orientation of the uniaxial indicatrix with the U-stage. With a mineral such as quartz which lacks visible crystallographic features such as cleavage, to obtain any information about the position of a-axes, it is necessary to use X-rays.

To determine the position of the optic axis (a symmetry axis of the indicatrix), we aim to set it parallel to A4, or if this is not possible to set it vertical. Proceed as follows:

(1) With all other axes at zero, rotate about A1 to an extinction position.

(2) A trial tilt is made about A2. If extinction is maintained, the correct extinction position under (1) above was chosen (but see *notes* below). If the grain illuminates on tilting, A2 is restored to zero, and A1 rotated to the alternative extinction position.

(3) A control tilt about A4 is made. The grain will normally show some illumination.

(4) With the control tilt on A4 maintained (say $30°$), extinction is restored by tilt about A2. The optic axis should now be either parallel to A4 or vertical, and extinction will be maintained on rotating A4.

(5) Return A4 to zero and rotate the microscope stage A5. If the c-axis is parallel to A4 the grain will illuminate, if the c-axis is vertical the grain will remain extinguished.

Notes: A complication arises if the c-axis is initially horizontal in the section. Proceeding under (1) and (2) above, it will be found that such a grain in either extinction position about A1 will remain extinguished when tilting A2. The correct position is chosen by first tilting A4, and then tilting

A2. If the grain remains extinguished the *c*-axis is N–S, and the alternative extinction position about A1 must be chosen.

With some experience, it is usually possible to anticipate whether the *c*-axis is to be set horizontal or vertical, and step (5) can be omitted. A grain with a near vertical *c*-axis has an interference colour which is initially low, and which becomes progressively lower near the extinction position reached with the U-stage. On the other hand, a grain with a near horizontal *c*-axis has initially a higher interference colour, and near the final extinction position reached with the U-stage, the interference colour actually increases in order.

If the grain remains extinguished during rotation of A1, it must not be assumed that the *c*-axis is vertical. Errors of up to 20° are made in this way. The correct position of A1 under (1) above must be found by making trial tilts about A2 in various positions of A1 until extinction is maintained.

It is worth noting that the common strain shadows of quartz are nearly always sub-parallel to the *c*-axis. It will save time therefore if they are set E–W when finding the correct extinction position about A1.

Plotting the optic-axis. Having set the optic axis horizontal or vertical, the A1 and A2 readings are taken. It is essential to note whether A2 has been tilted to the left (L) or to the right (R), and whether the optic axis is vertical. To plot a reading such as (A1) 20°, (A2) 40° (R) – horizontal, rotate the tracing paper until the index mark reads 20° on the perimeter of the net, and plot the optic axis 40° along the E–W diameter of the net from the east side (for the lower hemisphere of the stereogram). If the *c*-axis has been set vertical, say 40° (R), then it must be plotted 40° along the E–W diameter to the west from the centre of the net.

Determination of the orientation of the indicatrix in biaxial minerals

The *X*, *Y* and *Z* optical directions are all symmetry axes of the biaxial indicatrix. The aim of the following procedure is to set these axes in turn parallel to A4. Proceed as follows:

(1) Rotate about A1 to an extinction position.

(2) Tilt A4. The grain will normally illuminate. If so, restore extinction by tilting A2.

(3) Retaining the tilt of A2, restore A4 to zero.

(4) If the grain illuminates further, the procedure is repeated, first rotating A1 to extinction, tilting A4, restoring extinction by tilting A2 and returning A4 to zero, until extinction is maintained throughout the entire rotation of A4.

(5) To determine whether *X*, *Y* or *Z* has been set parallel to A4, rotate A5 to a 45° position.

(a) If it is *Y*, then the O.A.P. is vertical with A4 as its normal. On rotating A4, a rapid fall of interference colour will be seen with extinction if an optic axis can be brought into the vertical position. Sometimes both optic axes can be placed in a vertical position. For each optic axis note the reading of A4 making sure to note the direction of rotation.

(b) If no optic axis can be found, then X or Z has been set parallel to A4. After checking A4 is in the zero position, insert an accessory plate to determine whether the direction is fast (X) or slow (Z).

(6) Find the orientation of the other optical directions by rotating to an alternative extinction position with A1 and following the above procedure. Sometimes the three directions can be directly determined, but if not, the third direction can be plotted normal to the other two.

Plotting the results. The optical directions are plotted from readings of A1 and A2 in the same way as for the uniaxial optic axis. It should be checked that the three directions are at an angle of 90° from each other. The optic axes plot on the XZ plane. If only one optic axis can be directly determined, the other may be plotted at an equal angle along the O.A.P. from X or Z. To plot a reading for an optic axis, the tracing paper should be rotated so that Y lies on the E—W diameter of the net. If the A4 rotation was away from you, the reading should be plotted on the great circle to which Y is the pole, towards the north from the E—W diameter (lower hemisphere projection). The axis is plotted to the south if A4 was rotated towards you.

CHAPTER 5

Routine Laboratory Procedures

The following guide to procedure is designed to help the student develop an efficient and quick laboratory routine. The preceding chapter should be constantly referred to for technical details.

THIN-SECTION PROCEDURES

A

Many inexperienced students faced with a thin section, spend a considerable time either wondering where to start, or examining the most insignificant mineral in the section. They eventually find that there is no time to examine the abundant and important minerals! To help overcome this it is always advisable to hold the thin section against a sheet of white paper or a window before using the microscope. The position and percentage of minerals with differing colour, opacity, alteration and grain size can be noted, thus providing a proper perspective for what has to be done with the microscope. It must be realised, of course, that an area of the thin section that appears to be of the same material, may prove under the microscope to be composed of two or more minerals.

B

Colour, relief, grain size and shape enable an initial subdivision of the minerals present to be made. To do this, the thin section should be examined in plane-polarised light, usually with a low- or medium-power objective.

First, make a list of minerals grouped according to colour. Second, partially close the lower diaphragm, and examine the relief of every group. This may result in further subdivision. Third, the list should again be subdivided if there are marked differences in grain size or shape within any group.

It is recommended that the student makes this preliminary list of minerals by quickly scanning the entire thin section.

C

We now have some idea as to the *probable* number of minerals in the rock, and the next step is to examine each mineral in some detail. It is preferable to start with the most abundant mineral present.

If the mineral is opaque, refer immediately to Determinative Table X.

If the mineral is transparent, it is necessary to measure the following three properties before referring to the Determinative Tables.

(1) *Refractive index*. Use relief and Becke line to *estimate* this.

(2) *Birefringence (for anisotropic minerals)*. This can be measured with reasonable accuracy, provided several grains of random orientation are present; look for the grain with the highest interference colour.

(3) *Nature of the indicatrix*. Determine whether the mineral is isotropic, uniaxial +ve or −ve, biaxial +ve or −ve, and measure $2V$; look for the grain with the lowest interference colour.

The Determinative Tables can now be referred to. If the mineral is strongly coloured, it is often best to start with Determinative Table IV. Otherwise start with Tables V—IX in which minerals are grouped according to the nature of their indicatrices.

D

The above three properties are sometimes sufficient to identify a mineral. Even so, the mineral identification, or the short-list of possibilities sorted from the Tables should be checked by reference to the orientation diagrams of Chapter 7. The presence or absence of cleavages, values of extinction angles, and the position of fast and slow directions are features which nearly always allow a final identification to be made.

Cautionary notes

"Instant recognition" of minerals can hardly be avoided by the more experienced worker, and the procedure to be adopted in such cases will naturally be one that seeks confirmation rather than routine identification. Unfortunately, even experienced professionals do not always check their "instant determinations", and with inexperienced students, mistakes are as common as correct identifications. "Instant determinations", insufficiently checked, commonly result in the worker overlooking other minerals that are superficially similar to that instantly recognised.

For example, the following mistakes may result from the "instant recognition" of pyroxenes, hornblende, mica, and feldspar or quartz.

(1) *Pyroxene*. The two cleavages approximately at $90°$ in sections at right angles to the *c*-axis are so familiar, that a student will frequently not bother to examine the prismatic sections. The minerals olivine and epidote, which are common associates of pyroxene, may be overlooked in this way. Furthermore, in many igneous rocks, orthopyroxene occurs with augite, and since the properties of the two minerals overlap, the orthopyroxene may be over-

looked when present in small quantities. A number of $2V$ determinations on suitable sections may be necessary to establish the presence of the ortho-pyroxene.

(2) *Hornblende*. The pleochroism and two cleavages at 120° are so famil-iar that many students misidentify the (100) sections of hornblende. Such sections are not pleochroic and show no cleavage at all, and are commonly identified as another mineral (usually tourmaline). The interference figure obtained from such sections readily distinguishes hornblende from tourma-line.

(3) *Micas*. As with hornblende, sections of micas do not always display cleavages (nor do all sections of the coloured micas display pleochroism). Basal sections of biotite and muscovite frequently cause confusion, simply as a result of an unsystematic approach to mineral identification.

(4) *Feldspar and quartz*. It is incorrect to rely on the common alteration and twinning of feldspar to distinguish it from quartz. Characteristic as these features are, they are nevertheless not always present or visible. The more essential properties are those of R.I., birefringence, and the nature of the indicatrix. Quartz, for example, should always be confirmed by obtaining a uniaxial +ve interference figure.

It is important, therefore, to always check all essential properties, and to make a very thorough preliminary examination of the thin section.

LOOSE- OR CRUSHED-GRAIN MOUNT PROCEDURES

These mounts are usually of two types: (1) a sedimentary grain assem-blage in which the grain sizes, shapes and degree of roundness are to be observed as well as mineral types; and (2) a crushed grain of a single mineral, scraped from a rock to enable a quick optical determination to be made, obviating the need for a thin section. Refractive index may be measured with more precision but birefringence with less precision in grain mounts, and the following recommended methods and procedures differ slightly from those for thin sections.

A

A quick scan of the mount in plane-polarised light is important for sedi-mentary grain assemblages, where more than one mineral may be present. As for thin sections, a list of the probable number of minerals present should be made on the basis of colour, relief, grain size and shape.

B

An estimate of R.I., birefringence, and the nature of the indicatrix must then be made for each mineral.

(1) *Birefringence*. Because there is no standard thickness of loose grains, it is difficult to accurately measure birefringence. Furthermore, sedimentary grains are sometimes so thick that the interference colours cannot be easily determined. Some guide to thickness, provided the grains are reasonably equidimensional, may be made by measuring their diameter. Grains with a prominent crystallographic shape or cleavage will not provide the necessary random orientations for birefringence measurement. Despite these problems, it is worthwhile to make as good an *estimate* of birefringence as is possible.

(2) *Refractive index*. R.I. can be measured more accurately with loose grains than in thin section, since we may immerse the grains in any number of liquids of known R.I. We may also separately measure ω and ϵ, or α, β and γ (N.B. this may not be possible or worthwhile in routine work unless the birefringence is at least moderately high). To determine specific refractive indices we must obtain a grain oriented so that the direction to be measured lies parallel to the slide and to the polariser vibration direction. Suitably oriented grains can be identified as follows:

(a) *Uniaxial minerals.* ω, the ordinary ray, can be measured in any section of a uniaxial mineral. ω is the only direction present in sections that provide a centred uniaxial-cross interference figure. In other sections, to determine ω it is first of all necessary to know whether the mineral is +ve or —ve. ω is the fast direction in any section of a +ve mineral, the slow direction in any section of a —ve mineral. Alternatively, ω can be determined as the direction of constant R.I. in several sections of varying orientation, (and the sign indirectly established). ϵ can only be measured in grains with the *c*-axis horizontal. Such grains will show the highest interference colours for a particular thickness, and should provide a centred flash figure. Euhedral prismatic crystals will naturally lie with their *c*-axis (and hence ϵ) parallel to the slide.

(b) *Biaxial minerals.* β can be determined in any direction in a grain which lies with the circular section of the indicatrix parallel to the slide (i.e. a grain that provides an optic-axis figure). α and β, or γ and β, may be determined on grains providing bisectrix figures (see Chapter 4 on orientation diagrams on how to identify these directions). α and γ may be determined, but without precise accuracy, in grains showing the highest interference colour for a particular thickness (α being the fast direction, γ the slow direction).

It may be impossible, without special equipment, to measure every R.I. of a mineral. For example, mica, when crushed, forms a mass of flakes that always lie with their cleavage parallel to the slide. In this case it is impossible to measure α.

(3) *Nature of the indicatrix*. Sometimes the nature of the indicatrix is easily determined. In other cases it may be difficult to obtain accurate information. For example, a euhedral prismatic mineral will always lie on the slide with a preferred orientation; in many cases (e.g. all uniaxial minerals), only flash figures will be obtained. Even in these cases, some intelligent

guesses can be made. For example, consistently straight extinction suggests the mineral is either uniaxial, or biaxial and orthorhombic. If flash figures are always obtained on the prismatic sections, the mineral is positive if length slow and negative if length fast.

Once the above three properties have been determined, the Determinative Tables can be referred to. More emphasis should be placed on the accurate determination of refractive index made in grain mounts. As was the procedure for thin sections, a final identification is made by reference to the systematic descriptions of Chapter 7 where details of mineral(s) sorted from the Tables should be checked.

CHAPTER 6

Determinative Tables

In addition to the Michel-Lévy interference colour chart (Plate 1), the following determinative tables are provided to assist in the identification of minerals:

(I)   Minerals listed in order according to their R.I.
(II)   Minerals that commonly display anomalous interference colours.
(III)   Minerals listed in order according to their birefringence.
(IV)   Coloured minerals, subdivided according to the nature of their indica-
        trices.
(V)   Isotropic minerals.
(VI)   Uniaxial positive minerals.
(VII)   Uniaxial negative minerals.
(VIII) Biaxial positive minerals.
(IX)   Biaxial negative minerals.
(X)   Opaque minerals.

DETERMINATIVE TABLE I

List of minerals in order according to their lowest R.I.

Lower R.I.		Higher R.I.		Mineral	δ^*	Indicatrix
1.41	(n)	1.47	(n)	opal (72)	0	isotropic
1.433	(n)	1.44	(n)	fluorite (111)	0	isotropic
1.461	(n)	1.509	(n)	sodalite group (78)	0	isotropic
1.466	(α)	1.494	(γ)	carnallite (114)	M	$2V_z = 66°$
1.468–1.479	(α)	1.473–1.483	(γ)	tridymite (70)	VL-L	$2V_z = 30° - 90°$
1.468	(n)	1.512	(n)	allophane (49)	0	isotropic
1.470–1.500	(ϵ)	1.472–1.505	(ω)	chabazite group (87)	VL-L	uniax −ve, +ve
1.473–1.483	(α)	1.485–1.496	(γ)	natrolite (79)	L	$2V_z = 58° - 64°$
1.476–1.505	(α)	1.479–1.512	(γ)	heulandite (83)	VL-L	$2V_z = 0° - 74°$
1.479	(n)	1.493	(n)	analcite (85)	0	isotropic
1.480–1.600	(α)	1.500–1.640	(γ)	smectite group (46)	M-H	$2V_x = 0°$ −moderate
1.482–1.500	(α)	1.496–1.513	(γ)	stilbite (84)	L	$2V_x = 28° - 49°$
1.483–1.505	(α)	1.486–1.514	(γ)	phillipsite (89)	VL-L	$2V_z = 60° - 80°$
1.484	(ϵ)	1.487	(ω)	cristobalite (71)	VL	uniax −ve
1.485	(n)	1.620	(n)	volcanic glass (91)	0	isotropic
1.486	(ϵ)	1.658	(ω)	calcite − pure (124)	VH	uniax −ve
1.488–1.503	(ϵ)	1.490–1.528	(ω)	cancrinite (77)	VL-M	uniax −ve
1.490	(n)	1.490	(n)	sylvite (113)	0	isotropic
1.494	(α)	1.516	(γ)	kainite (116)	M	$2V_x = 85°$
1.497–1.530	(α)	1.518–1.544	(γ)	thomsonite (82)	L-M	$2V_z = 38° - 75°$
1.498	(α)	1.502	(γ)	wairakite (86)	VL	$2V_z = 70° -105°$
1.500	(ϵ)	1.679	(ω)	dolomite − pure (124)	VH	uniax −ve
1.502–1.514	(α)	1.514–1.526	(γ)	laumontite (88)	L	$2V_x = 26° - 47°$
1.504	(β)	1.508	(β)	mesolite (80)	VL	$2V_z =$ ca. $80°$
1.507–1.513	(α)	1.517–1.521	(γ)	scolecite (81)	L	$2V_x = 36° - 56°$
1.508–1.511	(ω)	1.509–1.511	(ϵ)	leucite (75)	0-VL	isotropic or uniax +ve
1.509	(ϵ)	1.700	(ω)	magnesite − pure (124)	VH	uniax −ve
1.510–1.548	(ϵ)	1.690–1.750	(ω)	ankerite (124)	VH	uniax −ve

	Mineral		Biref.	Optics
1.518—1.527 (α)	alkali-feldspar (73)	1.524—1.539 (γ)	L	$2V_x = 0°—98°$
1.519—1.521 (α)	gypsum (121)	1.529—1.531 (γ)	L	$2V_z = 58°$
1.520 (α)	kieserite (117)	1.584 (γ)	H	$2V_z = 55°$
1.52 (ϵ)	dahllite—francolite (125)	1.63 (ω)	VL-M	uniax -ve
1.522—1.560 (α)	cordierite (28)	1.527—1.578 (γ)	L-M	$2V_x = 42°—104°$
1.525—1.548 (α)	lepidolite (33)	1.551—1.587 (γ)	M-H	$2V_x = 0°—58°$
1.525—1.560 (α)	vermiculite (48)	1.545—1.585 (γ)	M	$2V_x$ small
1.526—1.542 (ϵ)	nepheline (76)	1.529—1.547 (ω)	VL-L	uniax -ve
1.526—1.544 (ω)	chalcedony (69)	1.532—1.553 (ϵ)	L	uniax +ve
1.526—1.563 (α)	kaolinite group (45)	1.556—1.571 (γ)	VL-L	$2V_x = 24°—50°$, $2V_z = 52°—80°$
1.526—1.575 (α)	plagioclase-feldspar (74)	1.536—1.588 (γ)	L	$2V_z = 73°—135°$
1.530—1.531 (α)	aragonite (123)	1.685—1.686 (γ)	VH	$2V_z = 18°$
1.530—1.625 (α)	biotite—phlogopite (34)	1.558—1.696 (γ)	M-VH	$2V_x = 0°—25°$
1.532—1.570 (α)	serpentine (37)	1.545—1.584 (γ)	VL-M	$2V_x = 20°—60°$
1.533—1.544 (ω)	apophyllite (44)	1.535—1.544 (ϵ)	0-VL	uniax +ve, -ve
1.534 (n)	langbeinite (115)	1.534 (n)	0	isotropic
1.534—1.556 (α)	pyrophyllite (38)	1.596—1.601 (γ)	H	$2V_x = 53°—62°$
1.539—1.550 (α)	talc (39)	1.584—1.596 (γ)	H	$2V_x = 0°—30°$
1.540—1.571 (ϵ)	scapolite (90)	1.546—1.600 (ω)	L-H	uniax -ve
1.543—1.634 (α)	stilpnomelane (40)	1.576—1.745 (γ)	H-VH	$2V_x = 0°—40°$
1.544 (n)	halite (112)	1.544 (n)	0	isotropic
1.544 (ω)	quartz (68)	1.553 (ϵ)	L	uniax +ve
1.545—1.630 (α)	illite (47)	1.570—1.670 (γ)	M-H	$2V_x$ small
1.547 (α)	polyhalite (118)	1.567 (γ)	M	$2V_x = 64°$
1.552—1.578 (α)	muscovite (31)	1.587—1.617 (γ)	H	$2V_x = 0°—47°$
1.559—1.590 (ω)	brucite (104)	1.579—1.600 (ϵ)	L-M	uniax +ve
1.562—1.630 (α)	Mg-chlorite (36)	1.565—1.630 (γ)	0-L	$2V_z = 0°—45°$
1.564—1.580 (α)	paragonite (32)	1.600—1.609 (γ)	M-H	$2V_x = 0°—46°$
1.564—1.600 (ϵ)	beryl (27)	1.568—1.608 (ω)	VL-L	uniax -ve
1.565—1.571 (α)	gibbsite (105)	1.580—1.595 (γ)	L-M	$2V_z <20°$
1.569—1.574 (α)	anhydrite (120)	1.609—1.618 (γ)	H	$2V_z = 42°—44°$
1.572 (ω)	alunite (122)	1.592 (ϵ)	M	uniax +ve

* Birefringence is abbreviated: 0 = zero; VL = 0.00—0.005; L = 0.005—0.015; M = 0.015—0.030; H = 0.030—0.065; VH = >0.065.

DETERMINATIVE TABLE I (continued)

Lower R.I.	Higher R.I.	Mineral	δ*	Indicatrix
1.585–1.616 (α)	1.600–1.644 (γ)	glauconite (35)	L-H	$2V_x = 0° - 20°$
1.592–1.643 (α)	1.621–1.674 (γ)	humite group (3)	M-H	$2V_z = 65° - 85°$
1.595–1.610 (α)	1.632–1.645 (γ)	pectolite (57)	H	$2V_z = 50° - 63°$
1.596–1.695 (α)	1.615–1.722 (γ)	anthophyllite–gedrite (59)	L-M	$2V_x = 70° - 120°$
1.597 (ϵ)	1.816 (ω)	rhodochrosite – pure (124)	VH	uniax –ve
1.599–1.688 (α)	1.622–1.705 (γ)	tremolite–Fe-actinolite (61)	M	$2V_x = 64° - 86°$
1.606–1.634 (α)	1.616–1.644 (γ)	topaz (10)	L	$2V_z = 48° - 70°$
1.606–1.702 (α)	1.627–1.718 (γ)	glaucophane–riebeckite (65)	L-M	$2V_x = 0° - 89°$
1.610–1.650 (ϵ)	1.635–1.675 (ω)	tourmaline (29)	M-H	uniax –ve
1.611–1.632 (α)	1.632–1.665 (γ)	prehnite (43)	M-H	$2V_z = 65° - 69°$
1.612–1.700 (α)	1.630–1.710 (γ)	eckermannite–arfvedsonite (66)	L-M	$2V_x = 0° - 80°$
1.613–1.705 (α)	1.632–1.731 (γ)	hornblende group (62)	L-M	$2V_x = 10° - 120°$
1.616–1.690 (ϵ)	1.624–1.70 (ω)	melilite (26)	0-L	uniax +ve, –ve
1.616–1.640 (α)	1.631–1.653 (γ)	wollastonite (56)	L	$2V_x = 38° - 60°$
1.622 (α)	1.631 (γ)	celestine (119)	L	$2V_z = 51°$
1.628–1.665 (ϵ)	1.632–1.668 (ω)	apatite (125)	VL-L	uniax –ve
1.629–1.642 (α)	1.638–1.653 (γ)	andalusite (5)	L	$2V_x = 73° - 86°$
1.630–1.638 (α)	1.644–1.650 (γ)	margarite (41)	L	$2V_x = 40° - 67°$
1.630–1.671 (α)	1.630–1.690 (γ)	Fe-chlorite (36)	0-L	$2V_x = 0° - 30°$
1.635 (ϵ)	1.875 (ω)	siderite – pure (124)	VH	uniax –ve
1.635–1.696 (α)	1.655–1.729 (γ)	cummingtonite–grunerite (60)	M-H	$2V_x = 84° - 115°$
1.635–1.732 (α)	1.670–1.775 (γ)	Mg-olivine (1)	H	$2V_x = 73° - 98°$
1.636 (α)	1.648 (γ)	baryte (119)	L	$2V_z = 37°$
1.639–1.654 (α)	1.650–1.674 (γ)	monticellite (2)	L-M	$2V_x = 69° - 82°$
1.640 (α)	1.680 (γ)	spurrite (16)	H	$2V_x = 40°$
1.640–1.670 (α)	1.651–1.690 (γ)	mullite (8)	L-M	$2V_z = 45° - 61°$
1.640–1.681 (α)	1.652–1.692 (γ)	jadeite (54)	L	$2V_z = 67° - 86°$
1.640–1.682 (α)	1.660–1.690 (γ)	kataphorite (67)	L-M	$2V_x = 0° - 50°$
1.643–1.649 (α)	1.655–1.663 (γ)	clintonite (42)	L	$2V_x = 2° - 40°$
1.646–1.650 (α)	1.661–1.662 (γ)	boehmite (107)	L	$2V_z$ moderate to large

1.648–1.663 (α)	1.662–1.679 (γ)	spodumene (55)	L-M	$2V_z = 55°-70°$
1.650–1.702 (α)	1.689–1.796 (γ)	oxyphornblende (63)	M-VH	$2V_x = 56°-88°$
1.651–1.769 (α)	1.658–1.788 (γ)	orthopyroxene (50)	L-M	$2V_x = 45°-128°$
1.654–1.661 (α)	1.673–1.683 (γ)	sillimanite (7)	M	$2V_z = 20°-30°$
1.655–1.686 (α)	1.685–1.723 (γ)	dumortierite (9)	L-H	$2V_x = 13°-63°$
1.659–1.693 (α)	1.668–1.704 (γ)	axinite (30)	L	$2V_x = 63°-80°$
1.659–1.743 (α)	1.688–1.772 (γ)	augite group (51)	M-H	$2V_z = 25°-83°$
1.665 (α)	1.685 (γ)	lawsonite (24)	M	$2V_z = 76°-87°$
1.665–1.728 (α)	1.683–1.754 (γ)	pumpellyite (25)	L-M	$2V_z = 10°-92°$, $2V_x = 0°-80°$
1.670–1.715 (α)	1.690–1.734 (γ)	clinozoisite (21)	L	$2V_z = 14°-90°$
1.675 (n)	1.734 (n)	hydrogrossular (4)	0-VL	isotropic
1.682–1.722 (α)	1.705–1.751 (γ)	pigeonite (52)	M	$2V_z = 0°-30°$
1.685–1.691 (α)	1.701–1.707 (γ)	barkevikite (64)	L-M	$2V_x = 40°-50°$
1.685–1.705 (α)	1.697–1.725 (γ)	zoisite (20)	VL-M	$2V_z = 0°-70°$
1.685–1.706 (α)	1.730–1.752 (γ)	diaspore (106)	H	$2V_z = 84°-86°$
1.690–1.813 (α)	1.706–1.891 (γ)	allanite (23)	L-VH	$2V_x = 40°-123°$
1.700–1.750 (α)	1.730–1.800 (γ)	aegirine-augite (53)	H	$2V_x = 70°-110°$
1.701–1.729 (α)	1.705–1.734 (γ)	sapphirine (15)	VL-L	$2V_x = 48°-114°$
1.701–1.736 (ε)	1.705–1.742 (ω)	vesuvianite (19)	VL-L	uniax -ve
1.702–1.710 (α)	1.718–1.726 (γ)	merwinite (18)	L-M	$2V_z = 52°-76°$
1.706–1.718 (α)	1.719–1.734 (γ)	kyanite (6)	L-M	$2V_x = 77°-82°$
1.707 (α)	1.730 (γ)	larnite (17)	M	$2V_z$ moderate
1.711–1.738 (α)	1.724–1.751 (γ)	rhodonite (58)	L	$2V_z = 61°-76°$
1.712–1.730 (α)	1.717–1.734 (γ)	chloritoid (12)	L-M	$2V_x = 40°-125°$
1.714–1.751 (α)	1.730–1.797 (γ)	epidote (21)	M-H	$2V_x = 64°-90°$
1.715 (n)	1.98 (n)	spinel (100)	0	isotropic
1.720–1.724 (ω)	1.810–1.828 (ε)	xenotime (127)	VH	uniax +ve
1.720 (n)	2.00 (n)	garnet (4)	0	isotropic
1.732–1.794 (α)	1.762–1.829 (γ)	piemontite (22)	M-VH	$2V_z = 64°-85°$
1.732–1.827 (α)	1.775–1.879 (γ)	Fe-olivine (1)	H	$2V_x = 46°-73°$
1.736 (n)	1.745 (n)	periclase (93)	0	isotropic
1.736–1.747 (α)	1.749–1.762 (γ)	staurolite (11)	L	$2V_z = 80°-92°$

* Birefringence is abbreviated: 0 = zero; VL = 0.00–0.005; L = 0.005–0.015; M = 0.015–0.030; H = 0.030–0.065; VH = >0.065.

DETERMINATIVE TABLE I (continued)

Lower R.I.	Higher R.I.	Mineral	δ*	Indicatrix
1.750–1.776 (α)	1.800–1.836 (γ)	aegirine (53)	H	$2V_x = 60° - 70°$
1.759–1.763 (ϵ)	1.765–1.772 (ω)	corundum (94)	L	uniax −ve
1.770–1.800 (α)	1.825–1.850 (γ)	monazite (126)	H-VH	$2V_z = 3° - 19°$
1.840–1.950 (α)	1.943–2.110 (γ)	sphene (13)	VH	$2V_z = 20° - 56°$
1.92 –1.96 (ω)	1.96 –2.02 (ϵ)	zircon (14)	H	uniax +ve
1.94 (α)	2.51 (γ)	lepidocrocite (103)	VH	$2V_x = 83°$
1.990–2.010 (ω)	2.093–2.100 (ϵ)	cassiterite (95)	VH	uniax +ve
2.0 (n)	2.0 (n)	limonite (103)	0	isotropic
2.00 (n)	2.16 (n)	chromite (100)	0	isotropic
2.217–2.275 (α)	2.356–2.415 (γ)	goethite (103)	VH	$2V_x = 0° - 27°$
2.30 (n)	2.38 (n)	perovskite (99)	0-VL	isotropic or $2V = 90°$
2.488 (ϵ)	2.561 (ω)	anatase (97)	VH	uniax −ve
2.583 (α)	2.700–2.741 (γ)	brookite (98)	VH	$2V_z = 0° - 30°$
2.605–2.616 (ω)	2.899–2.903 (ϵ)	rutile (96)	VH	uniax +ve
2.87 –2.94 (ϵ)	3.15 –3.22 (ω)	hematite (102)	VH	uniax −ve

* Birefringence is abbreviated: 0 = zero; VL = 0.00–0.005; L = 0.005–0.015; M = 0.015–0.030; H = 0.030–0.065; VH = >0.065.

DETERMINATIVE TABLE II

Minerals that commonly display anomalous interference colours

---

Apophyllite (44)
Brucite (104)
Chlorite (36)
Chloritoid (12)
Clinozoisite—epidote (21)
Melilite (26)
Na-rich amphiboles
Vesuvianite (19)
Zoisite (20)

DETERMINATIVE TABLE III

List of minerals in order according to their birefringence

δ	Mineral	R.I. range	Indicatrix
0.00 −0.001	analcite (85)	1.479−1.493	(isotropic)
0.00 −0.001	leucite (75)	1.508−1.511	uniax +ve
0.00 −0.002	apophyllite (44)	1.533−1.544	uniax +ve, −ve
0.00 −0.002	Ca-garnet (4)	1.675−2.0	(isotropic)
0.00 −0.002	perovskite (99)	2.30 −2.38	$2V$ = ca. 90° or isotropic
0.00 −0.013	melilite (26)	1.616−1.70	uniax +ve, −ve
0.00 −0.015	chlorite (36)	1.562−1.690	$2V_z$ = 0°−45°, $2V_x$ = 0°−30°
0.00 −0.016	dahllite−francolite (125)	1.52 −1.63	uniax −ve
0.001	mesolite (80)	1.504−1.508	$2V_z$ = ca. 80°
0.001−0.008	vesuvianite (19)	1.701−1.742	uniax −ve
0.001−0.009	kaolinite group (45)	1.526−1.571	$2V_x$ = 24°−50°, $2V_z$ = 52°−80°
0.002−0.007	tridymite (70)	1.468−1.483	$2V_z$ = 30°−90°
0.002−0.008	heulandite (83)	1.476−1.512	$2V_z$ = 0°−74°
0.002−0.008	apatite (125)	1.628−1.668	uniax −ve
0.002−0.015	chabazite group (87)	1.470−1.505	uniax −ve, +ve
0.002−0.025	cancrinite (77)	1.488−1.528	uniax −ve
0.003 ca.	cristobalite (71)	1.484−1.487	uniax −ve
0.003−0.007	nepheline (76)	1.526−1.547	uniax −ve
0.003−0.010	phillipsite (89)	1.483−1.514	$2V_z$ = 60°− 80°
0.004	wairakite (86)	1.498−1.502	$2V_z$ = 70°−105°
0.004−0.009	beryl (27)	1.564−1.608	uniax −ve
0.004−0.010	sapphirine (15)	1.701−1.734	$2V_x$ = 48°−114°
0.004−0.017	serpentine (37)	1.532−1.584	$2V_x$ = 20°− 60°
0.004−0.022	zoisite (20)	1.685−1.725	$2V_z$ = 0°− 70°
0.005−0.009	chalcedony (69)	1.526−1.553	uniax +ve
0.005−0.009	corundum (94)	1.759−1.772	uniax −ve
0.005−0.010	alkali-feldspar (73)	1.518−1.539	$2V_x$ = 0°− 98°
0.005−0.015	clinozoisite (21)	1.670−1.734	$2V_z$ = 14°− 90°

0.005–0.018	cordierite (28)	1.522–1.578	$2V_x = 42°—104°$
0.005–0.020	eckermannite—arfvedsonite (66)	1.612–1.710	$2V_x = 0°— 80°$
0.006–0.016	thomsonite (82)	1.497–1.544	$2V_z = 38°— 75°$
0.006–0.022	chloritoid (12)	1.712–1.740	$2V_z = 40°—125°$
0.006–0.022	glaucophane—riebeckite (65)	1.606–1.718	$2V_x = 0°— 89°$
0.006–0.036	scapolite (90)	1.540–1.600	uniax—ve
0.007–0.010	scolecite (81)	1.507–1.521	$2V_x = 36°— 56°$
0.007–0.011	topaz (10)	1.606–1.644	$2V_z = 48°— 70°$
0.007–0.013	plagioclase-feldspar (74)	1.526–1.588	$2V_z = 73°—135°$
0.007–0.014	axinite (30)	1.659–1.704	$2V_x = 63°— 80°$
0.007–0.019	orthopyroxene (50)	1.651–1.788	$2V_x = 45°—128°$
0.007–0.021	kataphorite (67)	1.640–1.690	$2V_x = 0°— 50°$
0.008–0.014	stilbite (84)	1.482–1.513	$2V_x = 28°— 49°$
0.008–0.015	jadeite (54)	1.640–1.692	$2V_z = 67°— 86°$
0.008–0.023	merwinite (18)	1.702–1.726	$2V_z = 52°— 76°$
0.009	quartz (68)	1.544–1.553	uniax +ve
0.009	celestine (119)	1.622–1.631	$2V_z = 51°$
0.009–0.011	andalusite (5)	1.629–1.653	$2V_x = 73°— 86°$
0.010	gypsum (121)	1.519–1.531	$2V_z = 58°$
0.010–0.015	laumontite (88)	1.502–1.526	$2V_x = 26°— 47°$
0.010–0.015	staurolite (11)	1.736–1.762	$2V_z = 80°— 92°$
0.010–0.021	brucite (104)	1.559–1.600	uniax +ve
0.010–0.028	pumpellyite (25)	1.665–1.754	$2V_z = 10°— 92°, 2V_x = 0°—80°$
0.010–0.037	dumortierite (9)	1.655–1.723	$2V_x = 13°— 63°$
0.011–0.014	rhodonite (58)	1.711–1.751	$2V_x = 61°— 76°$
0.011–0.020	monticellite (2)	1.639–1.674	$2V_x = 69°— 82°$
0.012	baryte (119)	1.636–1.648	$2V_z = 37°$
0.012 ca.	natrolite (79)	1.473–1.496	$2V_z = 58°— 64°$
0.012–0.014	clintonite (42)	1.643–1.663	$2V_x = 2°— 40°$
0.012–0.014	margarite (41)	1.630–1.650	$2V_x = 40°— 67°$
0.012–0.015	boehmite (107)	1.646–1.662	$2V_z$ moderate to large
0.012–0.016	kyanite (6)	1.706–1.734	$2V_x = 77°— 82°$
0.012–0.028	mullite (8)	1.640–1.690	$2V_z = 45°— 61°$

DETERMINATIVE TABLE III (continued)

δ	Mineral	R.I. range	Indicatrix
0.013–0.015	wollastonite (56)	1.616–1.653	$2V_x = 38°– 60°$
0.013–0.028	anthophyllite—gedrite (59)	1.596–1.722	$2V_x = 70°–120°$
0.013–0.078	allanite (23)	1.690–1.891	$2V_x = 40°–123°$
0.014–0.018	barkevikite (64)	1.685–1.707	$2V_x = 40°– 50°$
0.014–0.026	common hornblende (62)	1.618–1.700	$2V_x = 55°– 86°$
0.014–0.027	spodumene (55)	1.648–1.679	$2V_z = 55°– 70°$
0.014–0.029	hornblende group (62)	1.612–1.731	$2V_x = 10°–120°$
0.014–0.030	gibbsite (105)	1.565–1.595	$2V_z < 20°$
0.014–0.032	glauconite (35)	1.585–1.644	$2V_x = 0°– 20°$
0.015–0.049	epidote (21)	1.714–1.797	$2V_x = 64°– 90°$
0.017–0.027	tremolite—Fe-actinolite (61)	1.599–1.705	$2V_x = 64°– 86°$
0.017–0.035	tourmaline (29)	1.610–1.675	uniax −ve
0.018–0.033	augite group (51)	1.659–1.772	$2V_z = 25°– 83°$
0.018–0.039	lepidolite (33)	1.525–1.587	$2V_x = 0°– 58°$
0.019–0.024	sillimanite (7)	1.654–1.683	$2V_z = 20°– 30°$
0.020	polyhalite (118)	1.547–1.567	$2V_x = 64°$
0.020	alunite (122)	1.572–1.592	uniax +ve
0.020	lawsonite (24)	1.665–1.685	$2V_z = 76°– 87°$
0.020–0.030	vermiculite (48)	1.525–1.585	$2V_x$ small
0.020–0.040	smectite group (46)	1.480–1.640	$2V_x = 0°$–moderate
0.020–0.045	cummingtonite—grunerite (60)	1.635–1.729	$2V_x = 84°–115°$
0.020–0.094	oxyhornblende (63)	1.650–1.796	$2V_x = 56°– 88°$
0.021–0.035	prehnite (43)	1.611–1.665	$2V_z = 65°– 69°$
0.022	kainite (116)	1.494–1.516	$2V_x = 85°$
0.022–0.041	humite group (3)	1.592–1.674	$2V_z = 65°– 85°$
0.022–0.055	illite (47)	1.545–1.670	$2V_x$ small
0.023	larnite (17)	1.707–1.730	$2V_z$ moderate
0.023–0.029	pigeonite (52)	1.682–1.751	$2V_z = 0°– 30°$
0.025–0.088	piemontite (22)	1.732–1.829	$2V_z = 64°– 85°$

0.028	carnallite (114)	1.466—1.494	$2V_z = 66°$
0.028—0.038	paragonite (32)	1.564—1.609	$2V_x = 0° — 46°$
0.028—0.078	biotite—phlogopite (34)	1.530—1.696	$2V_x = 0° — 25°$
0.030—0.038	pectolite (57)	1.595—1.645	$2V_z = 50° — 63°$
0.030—0.050	aegirine-augite (53)	1.700—1.800	$2V_x = 70° —110°$
0.033—0.111	stilpnomelane (40)	1.543—1.745	$2V_x = 0° — 40°$
0.035—0.052	olivine (1)	1.635—1.879	$2V_x = 46° — 98°$
0.036—0.049	muscovite (31)	1.552—1.617	$2V_x = 0° — 47°$
0.039—0.050	talc (39)	1.539—1.596	$2V_x = 0° — 30°$
0.040	spurrite (16)	1.640—1.680	$2V_x = 40°$
0.040—0.047	anhydrite (120)	1.569—1.618	$2V_z = 42° — 44°$
0.040—0.060	zircon (14)	1.92 —2.02	uniax +ve
0.045—0.050	diaspore (106)	1.685—1.752	$2V_z = 84° — 86°$
0.045—0.062	pyrophyllite (38)	1.534—1.601	$2V_x = 53° — 62°$
0.045—0.075	monazite (126)	1.770—1.850	$2V_z = 3° — 19°$
0.050—0.060	aegirine (53)	1.750—1.836	$2V_x = 60° — 70°$
0.064	kieserite (117)	1.520—1.584	$2V_z = 55°$
0.073	anatase (97)	2.488—2.561	uniax —ve
0.086—0.107	xenotime (127)	1.720—1.828	uniax +ve
0.090—0.103	cassiterite (95)	1.990—2.100	uniax +ve
0.100—0.192	sphene (13)	1.840—2.110	$2V_z = 20° — 56°$
0.117—0.158	brookite (98)	2.583—2.741	$2V_z = 0° — 30°$
0.139—0.140	goethite (103)	2.217—2.415	$2V_x = 0° — 27°$
0.155	aragonite (123)	1.530—1.686	$2V_x = 18°$
0.172	calcite (124)	1.486—1.658	uniax —ve
0.179	dolomite (124)	1.500—1.679	uniax —ve
0.182—0.202	ankerite (124)	1.510—1.750	uniax —ve
0.191	magnesite (124)	1.509—1.700	uniax —ve
0.219	rhodochrosite (124)	1.597—1.816	uniax —ve
0.242	siderite (124)	1.635—1.875	uniax —ve
0.28	hematite (102)	2.87 —3.22	uniax —ve
0.286—0.294	rutile (96)	2.605—2.903	uniax +ve
0.57	lepidocrocite (103)	1.94 —2.51	$2V_x = 83°$

DETERMINATIVE TABLE IV

Coloured minerals in thin section (examples in brackets may appear to have the indicatrix indicated)

	Pink	Red	Violet-purple
Isotropic	garnet (4) sodalite group (78) spinel (100) volcanic glass (91)	volcanic glass (91)	fluorite (111)
Uniaxial +ve		cassiterite (95) rutile (96)	[Cr-chlorite (36)]
Uniaxial —ve	tourmaline (29) corundum (94) [Cr-chlorite (36)]	hematite (102) [biotite (34)]	[Cr-chlorite (36)]
Biaxial +ve	sphene (13) sapphirine (15) zoisite (20) piemontite (22) rhodonite (58) diaspore (106) Cr-chlorite (36) *Pyroxenes:* orthopyroxene (50) augite (51) pigeonite (52)	allanite (23) iddingsite (1)	piemontite (22) augite (51) Cr-chlorite (36)
Biaxial —ve	andalusite (5) dumortierite (9) sapphirine (15) Cr-chlorite (36) orthopyroxene (50)	biotite (34) allanite (23) iddingsite (1) *Amphiboles:* oxyhornblende (63) barkevikite (64) kataphorite (67)	dumortierite (9) axinite (30) Cr-chlorite (36)

Blue	Green		
sodalite group (78) spinel (100) fluorite (111)	garnet (4) [chlorite (36)] perovskite (99) spinel (100)		Isotropic
	vesuvianite (19) [chlorite (36)]		Uniaxial +ve
tourmaline (29) corundum (94) anatase (97)	vesuvianite (19) tourmaline (29) [biotite—phlogopite (34)] [chlorite (36)] corundum (94)		Uniaxial —ve
chloritoid (12) sapphirine (15) lawsonite (24)	chloritoid (12) sapphirine (15) pumpellyite (25) chlorite (36) perovskite (99) *Amphiboles:* anthophyllite—gedrite (59) cummingtonite (60) hornblende (62)	*Pyroxenes:* orthopyroxene (50) hedenbergite (51) augite (51) omphacite (51) fassaite (51) pigeonite (52) aegirine-augite (53)	Biaxial +ve
dumortierite (9) chloritoid (12) sapphirine (15) axinite (30) glaucophane—riebeckite (65)	bowlingite (1) andalusite (5) chloritoid (12) sapphirine (15) epidote (21) pumpellyite (25) *Sheet-silicates:* fuchsite (31) biotite—phlogopite (34) glauconite (35) chlorite (36) serpentine (37) stilpnomelane (40) clintonite (42) nontronite (46) vermiculite (48)	*Pyroxenes:* orthopyroxene (50) aegirine (53) aegirine-augite (53) *Amphiboles:* anthophyllite (59) actinolite (61) grunerite (60) hornblende (62) oxyhornblende (63) glaucophane (65) kataphorite (67)	Biaxial —ve

DETERMINATIVE TABLE IV (continued)

	Yellow	Brown
Isotropic	garnet (4) [melilite (26)] limonite (103) collophane (125)	garnet (4) opal (72) perovskite (99) chromite (100) limonite (103) collophane (125) volcanic glass (91)
Uniaxial +ve	melilite (26) cassiterite (95) rutile (96) xenotime (127)	zircon (14) vesuvianite (19) xenotime (127) cassiterite (95) rutile (96)
Uniaxial —ve	melilite (26) tourmaline (29) [biotite—phlogopite (34)] [chlorite (36)] corundum (94) anatase (97) dahllite—francolite (125)	vesuvianite (19) tourmaline (29) [biotite—phlogopite (34)] anatase (97) dahllite—francolite (125)
Biaxial +ve	humite group (3) staurolite (11) chloritoid (12) sapphirine (15) zoisite (20) clinozoisite (21) lawsonite (24) orthopyroxene (50) brookite (98) monazite (126)	sphene (13) allanite (23) pumpellyite (25) perovskite (99) brookite (98) gibbsite (105) diaspore (106) *Pyroxenes:* hedenbergite (51) augite (51)
Biaxial —ve	Fe-olivine (1) andalusite (5) staurolite (11) chloritoid (12) sapphirine (15) epidote (21) axinite (30) *Sheet-silicates:* biotite—phlogopite (34) chlorite (36) stilpnomelane (40) nontronite (46) orthopyroxene (50) goethite (103) lepidocrocite (103)	dumortierite (9) allanite (23) pumpellyite (25) goethite (103) lepidocrocite (103) *Sheet-silicates:* biotite—phlogopite (34) stilpnomelane (40) clintonite (42) nontronite (46) vermiculite (48) *Pyroxene:* acmite (53)

	Grey	
	sodalite group (78) perovskite (99) spinel (100) collophane (125) volcanic glass (91)	Isotropic
		Uniaxial +ve
	tourmaline (29) anatase (97) dahllite—francolite (125)	Uniaxial —ve
Amphiboles: anthophyllite—gedrite (59) cummingtonite (60) hornblende (62)	perovskite (99)	Biaxial +ve
Amphiboles: anthophyllite (59) grunerite (60) actinolite (61) hornblende (62) oxyhornblende (63) barkevikite (64) glaucophane (65) eckermannite—arfvedsonite (66) kataphorite (67)		Biaxial —ve

DETERMINATIVE TABLE V

Isotropic minerals in order according to their R.I.
(refer to Determinative Table III for minerals which are pseudo-isotropic with a zero or very weak birefringence)

R.I.	Mineral	Remarks
1.41 —1.47	opal (72)	organic, secondary
1.433—1.44	fluorite (111)	often blue or violet
1.461—1.509	sodalite group (78)	feldspathoid, six-sided crystals
1.468—1.512	allophane (49)	clay mineral
1.479—1.493	analcite (85)	zeolite
1.485—1.62	volcanic glass (91)	
1.490	sylvite (113)	water soluble, salty
1.508—1.511	leucite (75)	feldspathoid, often weak δ and twins
1.52 —1.63	collophane (125)	phosphate, yellow, brown
1.534	langbeinite (115)	water soluble
1.544	halite (112)	water soluble, salty
1.675—1.734	hydrogrossular (4)	garnet, often weak δ
1.715—1.98	spinel (100)	octahedral crystals
1.720—1.770	pyrope (4)	garnet in ultramafics, eclogites
1.735—1.770	grossularite (4)	garnet, often weak δ
1.736—1.745	periclase (93)	cubic cleavage, alters to brucite
1.770—1.820	almandine (4)	common metamorphic garnet
1.790—1.810	spessartine (4)	garnet, usually in pegmatites
1.850—2.0	andradite (4)	garnet, Ti-rich varieties dark-brown in section, often weak δ
1.86	uvarovite (4)	Cr-garnet, green
2.0	limonite (103)	yellow-brown, secondary
2.00 —2.16	chromite (100)	dark brown, in ultramafics
2.30 —2.38	perovskite (99)	often brown and weak δ

DETERMINATIVE TABLE VI

Uniaxial positive minerals in order according to their R.I.
(refer to Determinative Table VIII for pseudo-uniaxial minerals with $2V = 0°$ or small)

Mineral	ω	ϵ	δ
Chabazite group (87)	ca. 1.47 —1.50	ca. 1.47 —1.50	0.002—0.015
Leucite (75)	1.508—1.511	1.509—1.511	0.00 —0.001
Chalcedony (69)	1.526—1.544	1.532—1.553	0.005—0.009
Apophyllite (44)	1.533—1.544	1.535—1.544	0.00 —0.002
Quartz (68)	1.544	1.553	0.009
Brucite (104)	1.559—1.590	1.579—1.600	0.010—0.021
Alunite (122)	1.572	1.592	0.020
Melilite (26)	1.630—1.650	1.637—1.650	0.00 —0.008
Xenotime (127)	1.720—1.724	1.810—1.828	0.086—0.107
Zircon (14)	1.92 —1.96	1.96 —2.02	0.04 —0.06
Cassiterite (95)	1.990—2.010	2.093—2.100	0.090—0.103
Rutile (96)	2.605—2.616	2.899—2.903	0.286—0.294

DETERMINATIVE TABLE VII

Uniaxial negative minerals in order according to their R.I
(refer to Determinative Table IX for pseudo-uniaxial minerals with $2V = 0°$ or small)

Mineral	ω	ϵ	δ
Chabazite group (87)	1.472—1.505	1.470—1.500	0.002—0.015
Cristobalite (71)	1.487	1.484	0.003
Cancrinite (77)	1.490—1.528	1.488—1.503	0.002—0.025
Dahllite—fracolite (125)	(1.52—1.63)		0.00 —0.016
Nepheline (76)	1.529—1.547	1.526—1.542	0.003—0.007
Apophyllite (44)	1.544—1.5445	1.544	<0.001
Scapolite (90)	1.546—1.600	1.540—1.571	0.006—0.036
Beryl (27)	1.568—1.608	1.564—1.600	0.004—0.009
Melilite (26)	1.624—1.700	1.616—1.690	0.00 —0.013
Apatite (125)	1.632—1.668	1.628—1.665	0.002—0.008
Tourmaline (29)	1.635—1.675	1.610—1.650	0.017—0.035
Calcite — pure (124)	1.658	1.486	0.172
Dolomite—pure (124)	1.679	1.500	0.179
Ankerite (124)	ca. 1.69. —1.75	1.510—1.548	0.182—0.202
Magnesite — pure (124)	1.700	1.509	0.191
Vesuvianite (19)	1.705—1.742	1.701—1.736	0.001—0.008
Corundum (94)	1.765—1.772	1.759—1.763	0.005—0.009
Rhodochrosite — pure (124)	1.816	1.597	0.219
Siderite — pure (124)	1.875	1.635	0.242
Anatase (97)	2.561	2.488	0.073
Hematite (102)	3.15 —3.22	2.87 —2.94	0.28

DETERMINATIVE TABLE VIII

Biaxial positive minerals in order according to their R.I.

Mineral	α	β	γ
Carnallite (114)	1.466	1.475	1.494
Tridymite (70)	1.468—1.479	1.469—1.480	1.473—1.483
Natrolite (79)	1.473—1.483	1.476—1.486	1.485—1.496
Heulandite (83)	1.476—1.505	1.477—1.508	1.479—1.512
Phillipsite (89)	1.483—1.505	1.484—1.510	1.486—1.514
Thomsonite (82)	1.497—1.530	1.513—1.533	1.518—1.544
Wairakite (86)	1.498		1.502
Mesolite (80)		1.504—1.508	
Gypsum (121)	1.519—1.521	1.522—1.526	1.529—1.531
Kieserite (117)	1.520	1.533	1.584
Cordierite (28)	1.522—1.547	1.524—1.552	1.527—1.557
Plagioclase-feldspar (74)	1.526—1.575	1.532—1.584	1.536—1.588
Alkali-feldspar (73)	1.527—1.529	1.531—1.533	1.536—1.539
Dickite (45B)	1.560—1.562		1.566—1.571
Mg-chlorite (36)	1.562—1.630	1.562—1.630	1.565—1.630
Gibbsite (105)	1.565—1.571	ca. = α	1.580—1.595
Anhydrite (120)	1.569—1.574	1.574—1.579	1.609—1.618
Humite group (3)	1.592—1.643	1.602—1.653	1.621—1.674
Pectolite (57)	1.595—1.610	1.605—1.615	1.632—1.645
Topaz (10)	1.606—1.634	1.609—1.637	1.616—1.644
Prehnite (43)	1.611—1.632	1.615—1.642	1.632—1.665
Hornblende group (62)	1.613—1.650	1.618—1.660	1.635—1.670
Anthophyllite—gedrite (59)	1.62 —1.69	1.63 —1.705	1.645—1.715
Celestine (119)	1.622	1.624	1.631
Cummingtonite (60)	1.635—1.663	1.644—1.680	1.655—1.696
Olivine (1)	1.635—1.665	1.651—1.684	1.670—1.702
Baryte (119)	1.636	1.637	1.648
Mullite (8)	1.640—1.670	1.642—1.675	1.651—1.690
Jadeite (54)	1.640—1.681	1.645—1.684	1.652—1.692
Boehmite (107)	1.646—1.650		1.661—1.662
Spodumene (55)	1.648—1.663	1.655—1.669	1.662—1.679
Enstatite (50)	1.651—1.665	1.653—?	1.658—1.675
Sillimanite (7)	1.654—1.661	1.658—1.662	1.673—1.683
Augite group (51)	1.659—1.743	1.670—1.750	1.688—1.772
Lawsonite (24)	1.665	1.674	1.685
Pumpellyite (25)	1.665—1.710	1.670—1.730	1.683—1.730
Clinozoisite (21)	1.670—1.715	1.674—1.725	1.690—1.734
Pigeonite (52)	1.682—1.722	1.684—1.722	1.705—1.751

δ	$2V$	Remarks
0.028	66°	water soluble
0.002—0.007	30°—90°	SiO_2; platy crystals with twinning
ca. 0.012	58°—64°	a zeolite
0.002—0.008	0°—74°	a zeolite
0.003—0.010	60°—80°	a zeolite
0.006—0.016	38°—75°	a zeolite
0.004	70°—90°	a zeolite
0.001	ca. 80°	a zeolite
0.010	58°	evaporite or secondary mineral
0.064	55°	water soluble
0.005—0.016	76°—90°	sector twins, if present, are distinctive
0.07 —0.013	73°—90°	refer to Fig. 113 for $2V$ variation
0.009—0.010	82°—90°	
0.006—0.009	52°—80°	a clay mineral
0.00 —0.015	0°—45°	commonly green with anomalous brown int. colours
0.014—0.030	$<20^\circ$	in bauxites
0.040—0.047	42°—44°	evaporite or secondary mineral
0.022—0.041	65°—85°	usually yellow
0.030—0.038	50°—63°	usually in veins and cavities
0.007—0.011	48°—70°	
0.021—0.035	65°—69°	may resemble muscovite
0.018—0.024	60°—90°	an amphibole
0.013—0.025	60°—90°	an amphibole in metamorphics, straight extinction
0.009	51°	in evaporites and veins
0.020—0.035	65°—90°	a metamorphic amphibole, usually elongate
0.035—0.037	82°—90°	no good cleavage
0.012	37°	in veins and some sediments
0.012—0.028	45°—61°	in very high-temperature metamorphics
0.008—0.015	67°—86°	in very high-pressure metamorphics
0.012—0.015	moderate to large	in bauxites
0.014—0.027	55°—70°	Li-pyroxene in pegmatites
0.007—0.010	52°—90°	orthopyroxene
0.019—0.024	20°—30°	often fibrous, in high-grade metamorphics
0.018—0.033	25°—83°	clinopyroxenes
0.020	76°—87°	in high-pressure metamorphics
0.010—0.020	10°—90°	often green or brown; similar to epidote
0.005—0.015	14°—90°	anomalous interference colours
0.023—0.029	0°—30°	clinopyroxene in volcanics

DETERMINATIVE TABLE VIII (continued)

Mineral	α	β	γ
Zoisite (20)	1.685—1.705	1.688—1.710	1.697—1.725
Diaspore (106)	1.685—1.706	1.705—1.725	1.730—1.752
Allanite (23)	1.690—1.813	1.700—1.857	1.706—1.891
Aegirine-augite (53)	1.700—1.725	1.710—1.742	1.730—1.760
Sapphirine (15)	1.701—1.729	1.703—1.732	1.705—1.734
Merwinite (18)	1.702—1.710	1.710—1.718	1.718—1.726
Larnite (17)	1.707	1.715	1.730
Rhodonite (58)	1.711—1.738	1.716—1.741	1.724—1.751
Chloritoid (12)	1.712—1.730	1.717—1.734	1.719—1.740
Piemontite (22)	1.732—1.794	1.750—1.807	1.762—1.829
Staurolite (11)	1.736—1.747	1.741—1.754	1.749—1.762
Orthoferrosilite (50)	1.755—1.769	?—1.771	1.772—1.788
Monazite (126)	1.770—1.800	1.777—1.801	1.825—1.850
Sphene (13)	1.840—1.950	1.870—2.034	1.943—2.110
Perovskite (99)		2.30 —2.38	
Brookite (98)	2.583	2.584—2.586	2.700—2.741

δ	$2V$	Remarks
0.004—0.022	$0°—70°$	anomalous interference colours
0.045—0.050	$84°—86°$	in bauxites
0.013—0.078	$57°—90°$	dark brown or red-brown
0.030—0.035	$70°—90°$	a green clinopyroxene; alkaline igneous rocks
0.004—0.010	$66°—90°$	yellow, pink, or blue, in high-grade metamorphics
0.008—0.023	$52°—76°$	high-temperature Ca-silicate
0.023	moderate	high-temperature Ca-silicate
0.011—0.014	$61°—76°$	in Mn-rich deposits
0.006—0.022	$40°—90°$	platy crystals, similar to chlorite
0.025—0.088	$64°—85°$	pink or purple; similar to epidote
0.010—0.015	$80°—90°$	yellow, in medium-grade metamorphics
0.017—0.019	$60°—90°$	orthopyroxene
0.045—0.075	$3°—19°$	yellow, brown, or red; an accessory mineral
0.100—0.192	$20°—56°$	wedge-shaped or drop-like (sugary) brown crystals
0.00 —0.002	ca. $90°$	pseudocubic, often brown
0.117—0.158	$0°—30°$	yellow or brown; incomplete extinction

DETERMINATIVE TABLE IX

Biaxial negative minerals in order according to their R.I.

Mineral	α	β	γ
Smectite group (46)	1.48 —1.60		1.50 —1.64
Stilbite (84)	1.482—1.500	1.489—1.507	1.496—1.513
Kainite (116)	1.494	1.505	1.516
Wairakite (86)	1.498		1.502
Laumontite (88)	1.502—1.514	1.512—1.525	1.514—1.526
Scolecite (81)	1.507—1.513	1.516—1.520	1.517—1.521
Alkali-feldspar (73)	1.518—1.527	1.522—1.533	1.524—1.536
Cordierite (28)	1.522—1.560	1.524—1.574	1.527—1.578
Lepidolite (33)	1.525—1.548	1.548—1.585	1.551—1.587
Vermiculite (48)	1.525—1.560		1.545—1.585
Kaolinite group (45)	1.526—1.563		1.556—1.570
Plagioclase-feldspar (74)	1.526—1.575	1.532—1.584	1.536—1.588
Aragonite (123)	1.530—1.531	1.680—1.682	1.685—1.686
Biotite—phlogopite (34)	1.530—1.625	1.557—1.696	1.558—1.696
Serpentine (37)	1.532—1.570		1.545—1.584
Pyrophyllite (38)	1.534—1.556	1.586—1.589	1.596—1.601
Talc (39)	1.539—1.550	1.584—1.594	1.584—1.596
Stilpnomelane (40)	1.543—1.634	ca. = γ	1.576—1.745
Illite (47)	1.545—1.630		1.570—1.670
Muscovite (31)	1.552—1.578	1.582—1.615	1.587—1.617
Paragonite (32)	1.564—1.580	1.594—1.609	1.600—1.609
Glauconite (35)	1.585—1.616	ca. = γ	1.600—1.644
Anthophyllite (59)	1.596—1.62	1.605—1.63	1.615—1.645
Tremolite—Fe-actinolite (61)	1.599—1.688	1.612—1.697	1.622—1.705
Glaucophane—riebeckite (65)	1.606—1.702	1.622—1.712	1.627—1.718
Eckermannite—arfvedsonite (66)	1.612—1.700	1.625—1.709	1.630—1.710
Wollastonite (56)	1.616—1.640	1.627—1.650	1.631—1.653
Common hornblende (62)	1.618—1.680	1.628—1.691	1.636—1.700
Hornblende group (62)	1.618—1.705	1.628—1.729	1.636—1.731
Andalusite (5)	1.629—1.642	1.633—1.646	1.638—1.653
Margarite (41)	1.630—1.638	1.642—1.648	1.644—1.650
Fe-chlorite (36)	1.630—1.671	1.630—1.690	1.630—1.690
Monticellite (2)	1.639—1.654	1.646—1.664	1.650—1.674
Spurrite (16)	1.640	1.674	1.680
Kataphorite (67)	1.640—1.682	1.658—1.688	1.660—1.690
Clintonite (42)	1.643—1.649	1.655—1.662	1.655—1.663
Oxyhornblende (63)	1.650—1.702	1.682—1.769	1.689—1.796
Dumortierite (9)	1.655—1.686	1.667—1.722	1.685—1.723

δ	$2V$	Remarks
0.020—0.040	0°—moderate	clay minerals
0.008—0.014	28°—49°	a zeolite
0.022	85°	water soluble
0.004	75°—90°	a zeolite
0.010—0.015	26°—47°	a zeolite
0.007—0.010	36°—56°	a zeolite
0.005—0.009	0°—90°	
0.005—0.018	42°—90°	sector twins, if present, are distinctive
0.018—0.039	0°—58°	Li-mica
0.020—0.030	small	a clay mineral, often pseudomorphing biotite
0.001—0.007	24°—50°	clay minerals
0.007—0.013	45°—90°	refer to Fig. 113 for $2V$ variation
0.155	18°	$CaCO_3$; effervesces in dilute HCl
0.028—0.078	0°—25°	a mica; biotite is brown or green pleochroic
0.004—0.017	20°—60°	fine grained, flaky or fibrous
0.045—0.062	53°—62°	similar to colourless micas
0.039—0.050	0°—30°	soft, soapy feel, similar to colourless micas
0.033—0.111	0°—40°	similar to biotite, usually fine grained
0.022—0.055	small	a clay mineral
0.036—0.049	0°—47°	mica, usually colourless
0.028—0.038	0°—46°	very similar to muscovite
0.014—0.032	0°—20°	green pellets in marine sediments
0.013—0.025	70°—90°	a metamorphic amphibole; straight extinction
0.017—0.027	64°—86°	elongate amphibole in lower grade metamorphics
0.006—0.022	0°—89°	blue amphiboles; $2V$ of glaucophane $<50°$
0.005—0.020	0°—80°	amphibole in plutonic alkaline igneous rocks
0.013—0.015	38°—60°	usually in metamorphosed carbonate rocks
0.014—0.026	55°—86°	common amphibole
0.014—0.029	10°—90°	amphiboles
0.009—0.011	73°—86°	in low-pressure metamorphics
0.012—0.014	40°—67°	a colourless brittle-mica
0.00 —0.015	0°—30°	commonly green with anomalous blue int. colours
0.011—0.020	69°—82°	similar to olivine
0.040	40°	high-temperature Ca-silicate
0.007—0.021	0°—50°	an amphibole in alkaline igneous rocks
0.012—0.014	2°—40°	a brittle-mica; often green or brown
0.020—0.094	56°—88°	red-brown amphibole; phenocrysts in volcanics
0.010—0.037	13°—63°	usually blue, pleochroic; straight extinction

DETERMINATIVE TABLE IX (continued)

Mineral	α	β	γ
Axinite (30)	1.659—1.693	1.665—1.701	1.668—1.704
Grunerite (60)	1.663—1.696	1.680—1.708	1.696—1.729
Orthopyroxene (50)	1.665—1.755		1.675—1.772
Olivine (1)	1.665—1.827	1.684—1.869	1.702—1.879
Barkevikite (64)	1.685—1.691	1.696—1.700	1.701—1.707
Fe-gedrite (59)	1.690—1.695	1.705—1.710	1.715—1.722
Allanite (23)	1.690—1.813	1.700—1.857	1.706—1.891
Sapphirine (15)	1.701—1.729	1.703—1.732	1.705—1.734
Kyanite (6)	1.706—1.718	1.714—1.723	1.719—1.734
Pumpellyite (25)	1.710—1.728	1.730—1.748	1.730—1.754
Chloritoid (12)	1.712—1.730	1.717—1.734	1.719—1.740
Epidote (21)	1.714—1.751	1.721—1.784	1.730—1.797
Aegirine-augite (53)	1.725—1.750	1.742—1.780	1.760—1.800
Staurolite (11)	1.736—1.747	1.741—1.754	1.749—1.762
Aegirine (53)	1.750—1.776	1.780—1.820	1.800—1.836
Lepidocrocite (103)	1.94	2.20	2.51
Goethite (103)	2.217—2.275	2.346—2.409	2.356—2.415

δ	2V	Remarks
0.007—0.014	63°—80°	wedge-shaped crystals common
0.030—0.045	84°—90°	elongate metamorphic amphibole
0.010—0.017	45°—90°	
0.037—0.052	46°—90°	no good cleavage
0.014—0.018	40°—50°	red-brown amphibole; phenocrysts in alkaline rocks
0.025—0.028	82°—90°	elongate metamorphic amphibole
0.013—0.078	40°—90°	dark brown or red-brown
0.004—0.010	48°—90°	yellow, pink, or blue, in high-grade metamorphics
0.012—0.016	77°—82°	prominent cleavages in some sections
0.020—0.028	{ 0°—80° 88°—90°	often green or brown; similar to epidote
0.006—0.022	55°—90°	platy crystals; similar to chlorite
0.015—0.049	64°—90°	often a distinctive green; lack of 1st-order white
0.035—0.050	70°—90°	a green clinopyroxene; in alkaline rocks
0.010—0.015	88°—90°	yellow; in medium-grade metamorphics
0.050—0.060	60°—70°	a green clinopyroxene; in alkaline rocks
0.57	83°	FeO.OH; yellow-brown, pleochroic
0.139—0.140	0°—27°	FeO.OH; yellow or brown

DETERMINATIVE TABLE X

Opaque minerals

Colour in reflected light	Mineral	Remarks
Black, metallic lustre	graphite (92)	very soft, fine-grained or platy crystals
Black, metallic lustre	magnetite (100)	magnetic; octahedral crystals
Black, metallic lustre	hematite (102)	translucent red at thin edges; red streak
Black, metallic lustre	ilmenite (101)	non-magnetic; platy crystals; alters to leucoxene
White, like cotton-wool	leucoxene (101)	alteration product of Ti-rich minerals
Brownish-black, metallic or submetallic lustre	chromite (100)	translucent brown at thin edges
Brassy-yellow, metallic lustre	pyrite marcasite } (108)	
Bronze, metallic lustre	pyrrhotite (109)	magnetic
Golden-yellow, metallic lustre	chalcopyrite (110)	

CHAPTER 7

Mineral Descriptions

The following grouping of minerals according to structure and chemistry is adopted in this book:

A. Nesosilicates and sorosilicates

B. Cyclosilicates

C. Phyllosilicates

D. Inosilicates

E. Tectosilicates

F. Volcanic glass (not a mineral)

G. Non-silicates

For a straightforward discussion of the classes of silicate structures the reader is referred to Hurlbut (1971).

A grouping of minerals according to their optical properties is not followed because: (1) the range of properties such as R.I. and δ overlaps from mineral to mineral, (2) some do not lend themselves to a classification on the basis of optics — for example, a mineral may be +ve or —ve, (3) it is better to retain the natural grouping in mineral families such as the feldspathoids, members of which would be separated on the basis of their optical properties. Groupings according to optical properties are presented in the Determinative Tables (Chapter 6).

Each mineral is given a number in order of description. Generally every mineral is described separately, although the descriptions of some related types have been coordinated. Petrologically important and abundant minerals such as the feldspars, amphiboles and pyroxenes are described in greater detail than those of lesser importance.

For the majority of minerals, the description is laid out as follows: name and composition; crystallography; colour; (other) optical properties; orientation diagrams; occurrence; and distinguishing features. In addition to the numerical data provided, birefringence is described as very low (<0.005), low (0.005—0.015), moderate (0.015—0.030), high (0.030—0.065), and very high (>0.065), and relief (in canada balsam) as low (1.49—1.57), moderate (<1.49 and 1.57—1.68), high (1.68—1.78), very high (1.78—1.90) and extreme (>1.90). Where appropriate and relevant to detrital grain studies, properties such as colour and the orientation of cleavage fragments are noted.

All the important rock-forming minerals are described. The worker is referred to A. Winchell (1939, 1951) and H. Winchell (1965) for more comprehensive listings and tables.

A. NESOSILICATES AND SOROSILICATES

The nesosilicates contain isolated $[SiO_4]$ tetrahedral groups, and the soro-silicates paired tetrahedral groups $[Si_2O_7]$ (Hurlbut, 1971). Neso- and soro-silicates described here are: olivine; monticellite; the humite group; garnet; andalusite; kyanite; sillimanite; mullite; dumortierite; topaz; staurolite; chloritoid; sphene; zircon; sapphirine; spurrite; larnite; merwinite; vesuvianite; zoisite; clinozoisite—epidote; piemontite; allanite; lawsonite; pumpellyite; and melilite.

No. 1. OLIVINE (+ve and —ve) $(Mg,Fe)_2[SiO_4]$

A series from *forsterite* Mg_2SiO_4 to *fayalite* Fe_2SiO_4

Orthorhombic. Imperfect $\{010\}$ and poor $\{100\}$ cleavages. Euhedral stout crystals with $\{010\}$ $\{110\}$ $\{021\}$ and $\{001\}$ forms prominent. Also subhedral or granular.

Colour in thin section: usually colourless, but Fe-rich members may be pale yellow.

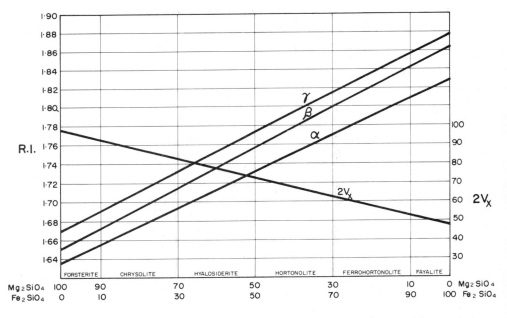

Fig. 50. Variation of R.I. and $2V_x$ with composition in olivine. Redrawn with permission from Poldervaart (1950, fig. 2).

Colour in detrital grains: usually a pale olive-green.

Optical properties: biaxial +ve or —ve. $2V_x$ = 98°—46°.
α = 1.635—1.827, β = 1.651—1.869, γ = 1.670—1.879.
δ = 0.035—0.052. $X = b$, $Y = c$, $Z = a$.
R.I. and δ increase, and $2V_x$ decreases with Fe content (Fig. 50). A useful estimate of composition can be made in thin section by estimating $2V$.

Orientation diagrams. *(100) section* (Fig. 51a): acute (pure forsterite) or obtuse bisectrix figure; δ' (moderate to high) = 0.016—0.042; slow along poorly developed cleavage; straight extinction; O.A.P. across cleavage.
(010) section (Fig. 51b): acute or obtuse (pure forsterite) bisectrix figure; δ' (low to moderate) = 0.010—0.019; fast along; no cleavage; O.A.P. across length of euhedral crystals.
(001) section (Fig. 51c): flash figure; δ (high) = 0.035—0.052; poor cleavage may be developed parallel to slow direction.

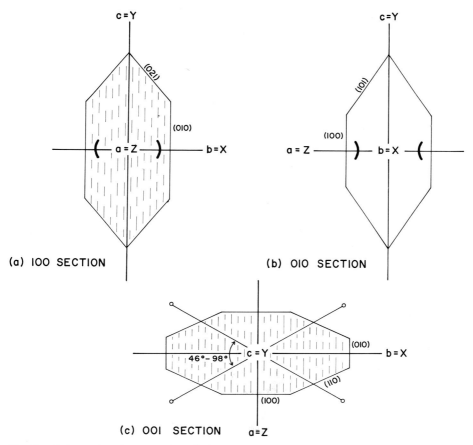

(a) 100 SECTION

(b) 010 SECTION

(c) 001 SECTION

Fig. 51. Orientation diagrams for olivine.

Occurrence. In a wide variety of igneous rocks, but most common in ultrabasic and basic types. Nearly pure forsterite occurs in dunite and peridotite, and moderately Mg-rich olivine in gabbro and basalt. Euhedral phenocrysts often display evidence of corrosion. Syenites and granites (and volcanic equivalents) may contain Fe-rich olivine. Almost pure forsterite occurs in thermally metamorphosed impure carbonate rocks. Fe-rich olivines are found in metamorphosed Fe-rich sediments. Olivine in metamorphosed gabbros may be surrounded by reaction rims (corona structures) usually composed of orthopyroxene and amphibole or garnet. Olivine is not a common detrital mineral since it is very susceptible to weathering and hydrothermal alteration. Alteration products (often as pseudomorphs) include serpentine, iddingsite, bowlingite, chlorite, talc, carbonates and iron oxides. Iddingsite and bowlingite are submicroscopic mixtures of iron oxides and chlorite (?). Iddingsite is blood-red, has a very high R.I. (ca. 1.76—1.89) and a high δ, and may be confused with biotite. Bowlingite is similar but green in colour.

Distinguishing features. Characterised by lack of colour, poor cleavage, high R.I. and δ, alteration products and occurrence. May be confused with diopside, augite, epidote, monticellite and humite. Distinguished from diopside and augite by lack of good cleavage, higher $2V$ and δ. Epidote has inclined extinction and anomalous interference colours. Monticellite and humite generally have a lower R.I. and/or δ.

No. 2. MONTICELLITE (—ve)   CaMg[SiO$_4$]

Fe may substitute for Mg, but most natural monticellites are very Mg-rich.

Orthorhombic. Imperfect {010} cleavage. Sometimes euhedral but often granular. Twinning with twin plane (031).

Colour in thin section: colourless.

Optical properties: biaxial —ve. $2V_x$ = 69°—82°.
α = 1.639—1.654, β = 1.646—1.664, γ = 1.650—1.674.
δ = 0.011—0.020. $X = b$, $Y = c$, $Z = a$.
Fe-rich varieties may have a higher R.I. and δ, and a smaller $2V$.

Orientation diagrams. Very similar to those for olivine, but always —ve and with a smaller δ.

Occurrence. In metamorphosed impure carbonate rocks. Also in some alkaline basic and ultrabasic rocks, often as parallel overgrowths on olivine.

Distinguishing features. Similar to olivine. Distinguished from olivine by its lower δ and from diopside and augite by its poor cleavage and —ve character.

No. 3. HUMITE GROUP (+ve)

Chondrodite	2Mg$_2$[SiO$_4$] · Mg(OH,F)$_2$
Humite	3Mg$_2$[SiO$_4$] · Mg(OH,F)$_2$
Clinohumite	4Mg$_2$[SiO$_4$] · Mg(OH,F)$_2$

Humite is orthorhombic. Chondrodite (β = 109°) and clinohumite (β = 100°) are monoclinic. Imperfect {100} cleavage. Subhedral or rounded anhedral crystals. Chondrodite,

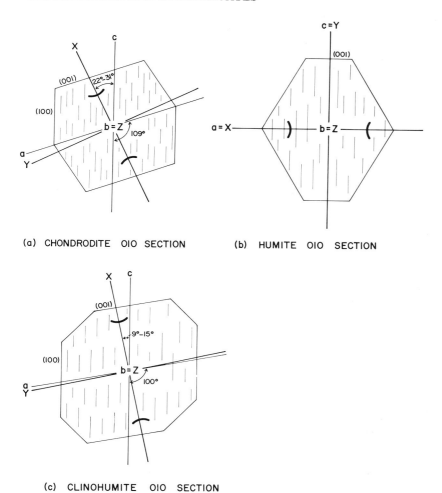

(a) CHONDRODITE OIO SECTION   (b) HUMITE OIO SECTION

(c) CLINOHUMITE OIO SECTION

Fig. 52. Orientation diagrams for the humite group.

humite and clinohumite may coexist in parallel intergrowths. Simple and lamellar twinning on {001} common in chondrodite and clinohumite.

Colour in thin section: pleochroic, colourless to pale yellow or pale to dark golden-yellow
with $X > Z > Y$.

Optical properties: biaxial +ve. $2V_z = 65°-85°$.
$\alpha = 1.592-1.643$, $\beta = 1.602-1.653$, $\gamma = 1.621-1.674$.
$\delta = 0.022-0.041$.
R.I. of clinohumite generally > humite > chondrodite, but ranges overlap. R.I. increases
with Fe substitution for Mg. Ti may raise R.I. in excess of 1.7.

Orientation diagrams. *(010) sections* (Fig. 52): acute bisectrix figure; δ' (low) = 0.010; humite extinguishes parallel to the poor $\{100\}$ cleavage, and the O.A.P. is across the cleavage; chondrodite and clinohumite have inclined extinction and the O.A.P. at an acute angle to the cleavage.
(001) sections: trace of the poor cleavage and straight extinction in all species; (001) section of humite displays the maximum δ with Z parallel to the cleavage.
(100) sections: no cleavage; (100) sections of chondrodite and clinohumite display the maximum δ.

Occurrence. In thermally metamorphosed and metasomatised carbonate rocks.

Distinguishing features. Pale-yellow colour, lack of good cleavage, moderate relief and moderate to high δ characteristic. The colourless species may be confused with olivine, but they usually have a smaller +ve $2V$. The yellow species may be confused with yellow tourmaline or staurolite. However, tourmaline is uniaxial, and staurolite has a different pleochroic scheme with $Z > X$, a higher R.I. and lower δ.

No. 4. GARNET GROUP (isotropic)

Pyralspite	pyrope	$Mg_3Al_2Si_3O_{12}$
	almandine	$Fe_3^{2+}Al_2Si_3O_{12}$
	spessartine	$Mn_3Al_2Si_3O_{12}$
Ugrandite	uvarovite	$Ca_3Cr_2Si_3O_{12}$
	grossularite	$Ca_3Al_2Si_3O_{12}$
	andradite	$Ca_3Fe_2^{3+}Si_3O_{12}$
Hydrogrossular		$Ca_3Al_2Si_2O_8(SiO_4)_{1-m}(OH)_{4m}$

Melanite and *schorlomite* are Ti-rich andradites.
There is continuous variation in composition within the two series pyralspite and ugrandite, but limited mixing between the two. The name given to any particular garnet is that of the dominant molecule present; pure end-members are rare.

Cubic. No cleavage. Euhedral dodecahedra (six-sided sections) and trapezohedra (eight-sided sections) common; also anhedral. Zoning is often present but may be visible only in the strongly coloured varieties. Complex twin patterns in some birefringent varieties.

Colour in thin section: colourless, pink or brown; less commonly yellow or green.

Colour in detrital grains: as above. but deeper colours, often red.

Optical properties: isotropic, though members of the ugrandite series, hydrogrossular, and spessartine may be weakly birefringent.
R.I. values for natural varieties with the following end-members dominant are: pyrope = 1.720–1.770; almandine = 1.770–1.820; spessartine = 1.790–1.810; uvarovite = ca. 1.86; grossularite = 1.735–1.770; andradite = 1.850–1.890 (Ti-rich andradite = ca. 1.86–2.0); hydrogrossular = 1.675–1.734.

Occurrence. Detrital grains of garnet are common in sediments. *Pyrope* occurs in some ultrabasic rocks, especially kimberlite. Eclogites and amphibolites contain a mixed pyrope–almandine garnet. *Almandine* is the typical garnet of regional metamorphism in pelitic and semi-pelitic schists, forming first in the upper part of the greenschist facies; earlier formed garnets usually contain significant spessartine. Less commonly found in thermally metamorphosed rocks and rare in igneous rocks. *Spessartine* is most common in granite-pegmatites. *Uvarovite* is very rare; found in some serpentinites and skarns. *Grossularite* is most common in metamorphosed carbonate rocks, and in metasomatic rocks, often with diopside. *Andradite* occurs in thermally metamorphosed impure carbonate rocks and skarns. *Melanite* and *schorlomite* are found in alkaline igneous rocks and skarns. *Hydrogrossular* occurs in metamorphosed and metasomatised carbonate rocks, in altered gabbros, and in rodingites with diopside.

Distinguishing features and identification of particular garnet species. The high relief, crystal shape, and isotropic character are distinctive. Garnet may be confused with spinel, but spinel crystallises as octahedra, and is commonly grey or green in thin section.

Individual species of garnet are not easy to determine. Various diagrams have been constructed to estimate the composition of a garnet from the three properties, R.I., density and cell size (Winchell, 1958) or from the two properties, R.I. and cell size (Sriramadas, 1957). Unfortunately, density measurements are not easy to make due to the common presence of inclusions in garnets; cell-size determination requires the use of X-rays. Neither of these methods nor chemical analysis provides satisfactory results for zoned garnets, and the best method at present is analysis using the X-ray microprobe. In some cases, R.I. by itself provides an estimate of composition. For example, hydrogrossular has the lowest R.I., andradite the highest. In addition, occurrence is a useful guide. The colour in thin section or hand specimen may be distinctive; whilst most garnets are red in hand specimen and colourless or pale pink in thin section, melanite and schorlomite are dark brown in thin section, uvarovite is green, and grossularite in hand specimen is often white or pale green. Members of the ugrandite series, hydrogrossular, and spessartine may exhibit a weak birefringence.

No. 5. ANDALUSITE (−ve) Al_2SiO_5

Small amounts of Fe or Mn may be present.

Orthorhombic. Distinct {110} and poor {100} cleavages. Usually euhedral elongate prismatic crystals with well developed {110} faces giving a square cross-section; rarely anhedral. The variety *chiastolite* contains graphitic inclusions, which, when viewed in cross-section, appear in the form of a cross running from the edges of the prism faces through the centre of the crystal (Fig. 54a).

Colour in thin section: colourless or pleochroic with X = pink, Y and Z = pale green or yellow.

Colour in detrital grains: more strongly coloured and pleochroic as above.

Optical properties: biaxial −ve. $2V_x = 73° - 86°$
$\alpha = 1.629 - 1.642$, $\beta = 1.633 - 1.646$, $\gamma = 1.638 - 1.653$.
$\delta = 0.009 - 0.011$. $X = c$, $Y = b$, $Z = a$.
R.I. increases with Fe and Mn. Some Mn-rich andalusites are +ve and have a higher R.I. than the range given above.

Orientation diagrams. *(100) section* (Fig. 53a): obtuse bisectrix figure; δ' (very low) = 0.004; fast along and straight extinction; cleavage is oblique to the section and may not be visible.
(010) section (Fig. 53b): flash figure; δ (low) up to 0.011; fast along and straight extinction; cleavage is oblique to the section and may not be visible.
(001) section (Fig. 53c): acute bisectrix figure; δ' (low) up to 0.006; symmetrical extinc-

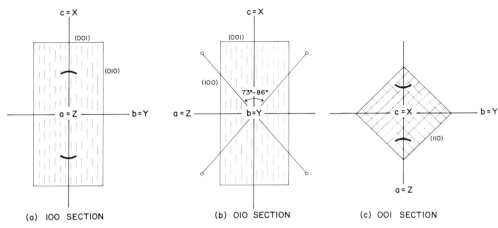

Fig. 53. Orientation diagrams for andalusite.

tion, two cleavages almost at right angles to each other; chiastolite displays inclusion cross parallel to Y and Z.

Occurrence. Typically developed in thermally metamorphosed pelitic rocks, but also in pelitic rocks regionally metamorphosed under high geothermal gradients. Commonly associated with cordierite. Andalusite changes to sillimanite at higher grades of metamorphism. Rarely occurs in granites and pegmatites. Frequently alters to sericite.

Distinguishing features. The shape of the crystals, the inclusion cross, the moderate relief, low δ, and occurrence are characteristic. The two cleavages and pleochroic scheme may cause confusion with orthopyroxene, but orthopyroxene is length slow.

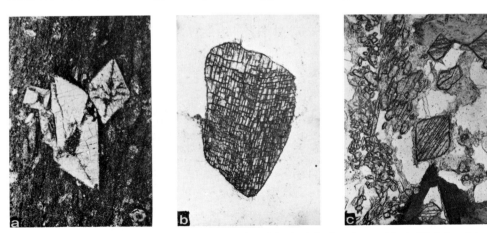

Fig. 54. The aluminium silicate polymorphs. (a) Andalusite (var. chiastolite) in slate, Berridale, N.S.W., Australia (view measures 3.2 mm × 2.0 mm). (b) Kyanite, in granulite (view measures 0.8 mm × 0.5 mm). (c) Sillimanite, in gneiss, Broken Hill, Australia (view measures 0.8 mm × 0.5 mm). Plane-polarised light.

No. 6. KYANITE (—ve) Al_2SiO_5

Triclinic, α = 90°5′, β = 101°, γ = 106°. Perfect {100} and less perfect {010} cleavages. Also {001} parting. Subhedral {100} tablets elongate parallel to c (blade-shaped crystals). Simple twins with {100} composition plane common; also multiple twins with {001} composition plane.

Colour in thin section: colourless.

Colour in detrital grains: blue, pleochroic with $Z > Y > X$.

Optical properties: biaxial —ve. $2V_x$ = 77°—82°.
α = 1.706—1.718, β = 1.714—1.723, γ = 1.719—1.734.
δ = 0.012—0.016. $X \wedge a$ = ca. 0°—5°, $Y \wedge b$ = ca. 30°, $Z \wedge c$ = ca. 30°.
Rare Cr-rich kyanites have been reported with γ up to 1.776.

Orientation diagrams. *(100) section (and cleavage fragments)* (Fig. 55a): acute bisectrix figure; δ' (low) = 0.006—0.011; Z onto good cleavage = ca. 30°; also (001) cross-parting visible.
(001) section (Fig. 55b): off-centred flash figure; δ' (low) = ca. 0.013; two well-developed cleavages at 74° to each other and a small extinction angle onto both; X is usually across the length of the crystals.
Optic-axis figures obtained from sections displaying no cleavage.
Maximum interference colours seen in sections showing (100) cleavage and nearly straight extinction (slow along).

Occurrence. Typically in high-grade regionally metamorphosed pelitic rocks. Also in granulites. More rarely in thermally metamorphosed rocks. Kyanite may change to sillimanite at higher grades of metamorphism. Also occurs in quartz segregation veins. Often conspicuous as detrital grains. Alters to sericite.

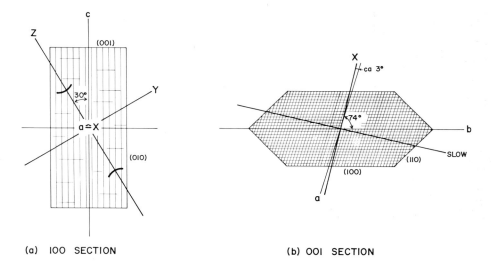

(a) 100 SECTION   (b) 001 SECTION

Fig. 55. Orientation diagrams for kyanite.

Distinguishing features. The high relief, low δ, very well-developed cleavages (Fig. 54b), and inclined extinction (especially on (100) sections) are distinctive. Not easily confused with other minerals. Detrital grains are characterised by their high relief, conspicuous cleavages, and pale-blue colour.

No. 7. SILLIMANITE (+ve) Al_2SiO_5

Orthorhombic. Perfect $\{010\}$ cleavage. Usually euhedral: stout prisms with a nearly square cross-section or commonly as minute fibres (variety *fibrolite*).

Colour in thin section: colourless.

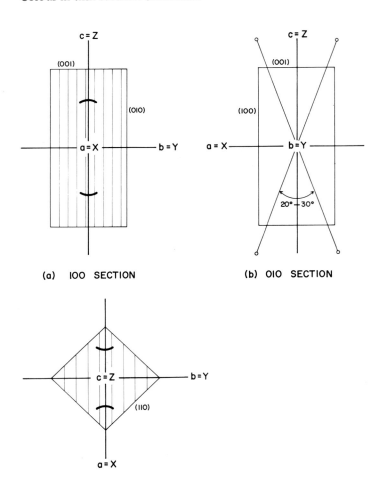

(a)   IOO SECTION

(b) OIO SECTION

(c) OOI   SECTION

Fig. 56. Orientation diagrams for sillimanite.

Colour in detrital grains: colourless or pleochroic with X = pale brown, Y = brown or green, Z = dark brown or blue.

Optical properties: biaxial +ve. $2V_z = 20°-30°$.
$\alpha = 1.654-1.661$, $\beta = 1.658-1.662$, $\gamma = 1.673-1.683$.
$\delta = 0.019-0.024$. $X = a$, $Y = b$, $Z = c$.

Orientation diagrams. *(100) section* (Fig. 56a): obtuse bisectrix figure; δ' (moderate) up to 0.021; slow along and straight extinction; cleavage rarely seen in small fibres.
(010) section (and cleavage fragments) (Fig. 56b): flash figure; δ (moderate) up to 0.024; slow along and straight extinction.
(001) section (Fig. 56c): acute bisectrix figure; δ' (low) up to 0.009; extinction positions often indistinct due to proximity of optic axes.
Optic-axis figures obtained in sections slightly oblique to (001).

Occurrence. A metamorphic mineral characteristic of high-grade schists, gneisses, migmatites, and hornfelses. Fibrolite masses are often embedded in micas and quartz.

Distinguishing features. The single cleavage in stout prisms or fibrous nature (Fig. 54c), the straight extinction, length-slow character, moderate δ, and occurrence are characteristic. Small fibres may be confused with anthophyllite or apatite. Apatite is fast along and has hexagonal cross-sections, though both these features are difficult to observe with small fibres embedded in other minerals. Anthophyllite has a more restricted occurrence, and usually a slightly lower R.I. It may be necessary to use X-rays to distinguish sillimanite from mullite.

No. 8. MULLITE (+ve) $3Al_2O_3 \cdot 2SiO_2$

Orthorhombic. Perfect $\{010\}$ cleavage. Long prisms with nearly square cross-sections, or fibrous.

Colour in thin section: colourless.

Optical properties: biaxial +ve. $2V_z = 45°-61°$.
$\alpha = 1.640-1.670$, $\beta = 1.642-1.675$, $\gamma = 1.651-1.690$.
$\delta = 0.012-0.028$. $X = a$, $Y = b$, $Z = c$.

Orientation diagrams. Almost identical to those for sillimanite.

Occurrence. Restricted to pelitic rocks thermally metamorphosed or fused (buchites) at very high temperatures; usually in xenoliths. A common refractory product.

Distinguishing features. Almost indistinguishable from sillimanite, but its occurrence is very restricted. X-ray tests may, with difficulty, distinguish the two.

No. 9. DUMORTIERITE (−ve) $(Al,Fe)_7O_3(BO_3)[SiO_4]_3$

Orthorhombic. Distinct $\{100\}$ and poor $\{110\}$ cleavages. Usually fibrous or bladed crystals elongate parallel to c. Sector twins with $\{110\}$ composition planes.

Colour in thin section: strongly pleochroic, usually blue or violet, also brown or pink, with $X > Y > Z$.

Optical properties: biaxial —ve. $2V_x = 13°-63°$.
$\alpha = 1.655-1.686$, $\beta = 1.667-1.722$, $\gamma = 1.685-1.723$.
$\delta = 0.010-0.037$. $X = c$, $Y = b$, $Z = a$.
Crystals have straight extinction, are length fast, and have the maximum absorption parallel to their length.

Occurrence. Usually associated with metasomatic or hydrothermal activity, and found in granite-pegmatites, gneisses, quartz veins, quartzites, and altered igneous rocks.

Distinguishing features. The strong pleochroism and blue colour are distinctive, together with the straight extinction and length-fast character. May be confused with tourmaline, but the maximum absorption of tourmaline is across the length of the crystals. Pale-coloured fine crystals of dumortierite may be confused with sillimanite, but sillimanite is length slow.

No. 10. TOPAZ (+ve) $Al_2[SiO_4](OH,F)_2$

Orthorhombic. Perfect $\{001\}$ cleavage. Sometimes euhedral prismatic crystals. Often anhedral.

Colour in thin section: colourless.

Colour in detrital grains: X and Y = yellow, Z = pink.

Optical properties: biaxial +ve. $2V_z = 48°-70°$.
$\alpha = 1.606-1.634$, $\beta = 1.609-1.637$, $\gamma = 1.616-1.644$.
$\delta = 0.007-0.011$. $X = a$, $Y = b$, $Z = c$.
$2V$ increases and R.I. decreases with increase in F.

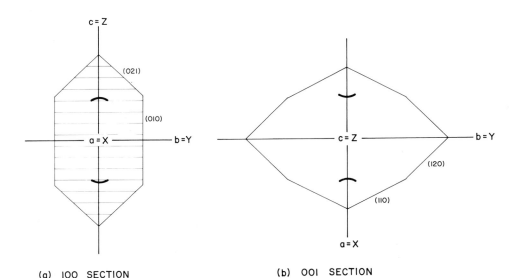

(a) 100 SECTION   (b) 001 SECTION

Fig. 57. Orientation diagrams for topaz.

Orientation diagrams. *(100) section* (Fig. 57a): obtuse bisectrix figure; δ' (low) = ca. 0.007; O.A.P. across cleavage; slow along length of prisms; straight extinction.
(010) section: similar appearance to (100) section but flash figure and maximum interference colours seen.
(001) section and cleavage fragments (Fig. 57b): acute bisectrix figure; δ' (very low) = ca. 0.003; no cleavage visible.
Optic-axis figures obtained in sections slightly oblique to (001).

Occurrence. Present in some granites and rhyolites, but most common in granite pegmatites and greisen associated with pneumatolytic activity. Alters to sericite and fluorite.

Distinguishing features. The moderate relief, low δ, and biaxial +ve character are distinctive. Apatite resembles topaz but is uniaxial −ve.

No. 11. STAUROLITE (+ve, rarely −ve) $\qquad\qquad Fe_2^{2+} Al_9 Si_4 O_{22}(O,OH)_2$

There may be minor substitutions of Mg for Fe^{2+}, and Fe^{3+} for Al.

Monoclinic, but pseudo-orthorhombic with β = ca. 90°. Imperfect $\{010\}$ cleavage. Euhedral or subhedral prismatic crystals with $\{110\}$ and $\{010\}$ dominating. Penetration twins, not usually seen in thin section.

Colour in thin section: pleochroic from colourless to yellow with $Z > Y > X$.

Colour in detrital grains: as above but deeper colours.

Optical properties: biaxial +ve or −ve. $2V_z$ = 80°−92°.
α = 1.736−1.747, β = 1.741−1.754, γ = 1.749−1.762.
δ = 0.010−0.015. $X = b$, $Y = a$, $Z = c$.

Orientation diagrams. *(100) section* (Fig. 58a): flash figure; δ (low) up to 0.015; slow parallel to weak cleavage; straight extinction.
(010) section (Fig. 58b): obtuse bisectrix figure; δ' (low) = ca. 0.007; straight extinction to prismatic edges; no cleavage.
(001) section (Fig. 58c): acute bisectrix figure; δ' (low) = ca. 0.006; O.A.P. across weak cleavage; straight extinction to cleavage.

Occurrence. Often as poikiloblasts in medium-grade regionally metamorphosed pelites. Often associated with chloritoid or kyanite. Kyanite sometimes replaces staurolite. May alter to sericite and chlorite, but is resistant to weathering and common as a detrital mineral.

Distinguishing features. High relief, low δ and yellow colour very distinctive. May be confused with yellow tourmaline, which however, is uniaxial and fast along.

No. 12. CHLORITOID (+ve, sometimes −ve)
$$(Fe^{2+},Mg,Mn)_2(Al,Fe^{3+})Al_3O_2[SiO_4]_2(OH)_4$$

Ottrelite is Mn-rich chloritoid.

Monoclinic or triclinic (β = 102°, γ = ca. 90°). Perfect $\{001\}$ and poor $\{110\}$? cleavages. Subhedral tablets parallel to $\{001\}$ with pseudohexagonal outlines. Also anhedral. Lamellar twinning with $\{001\}$ composition plane common. Hour-glass structure common.

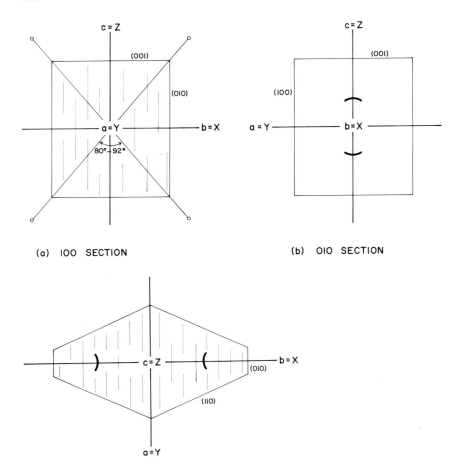

(a) 100 SECTION

(b) 010 SECTION

(c) 001 SECTION

Fig. 58. Orientation diagrams for staurolite.

Colour in thin section: colourless to green or blue-green pleochroic usually with X = green, Y = blue, Z = colourless or pale yellow.

Colour in detrital grains: as above but deeper colours.

Optical properties: biaxial +ve or —ve. $2V_z$ = $40°-125°$ (usually <$90°$).
α = 1.7 2—1.730, β = 1.717—1.734, γ = 1.719—1.740.
δ = 0 006—0.022. X = b (rarely Y = b), $Y \wedge a$ = $2°-30°$, $Z \wedge c$ = $14°-42°$ (Z onto $\perp$ 001 = $2°-30°$).
Interference colours may be anomalous.

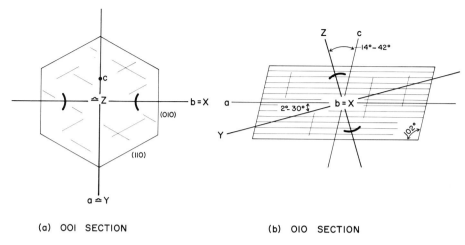

(a) OOI SECTION   (b) OIO SECTION

Fig. 59. Orientation diagrams for chloritoid.

Orientation diagrams. *(001) section and cleavage fragments* (Fig. 59a): acute bisectrix figure (slightly off-centred); δ' (low) = ca. 0.006; extinction positions often indistinct due to proximity of optic axes; poor $\{110\}$ cleavages rarely visible; very weak pleochroism. *(010) section* (Fig. 59b): obtuse bisectrix figure; δ' (low) = ca. 0.005; inclined extinction, fast onto perfect cleavage; twinning commonly observed; poor cleavages // c rarely visible. *(100) section*: similar to (010) section, but flash figure, straight extinction and displays maximum interference colours.

Occurrence. In low-grade regionally metamorphosed pelitic rocks, often as small post-tectonic porphyroblasts (usually monoclinic). Rarely in quartz veins (usually triclinic).

Distinguishing features. High relief, low δ, green colour, and platy crystals with twinning distinctive. May be confused with chlorite, but chlorite has a lower R.I., and usually a smaller $2V$ and extinction angle. The brittle-micas are optically −ve. Chloritoid does not display the mottled extinction many other platy minerals do.

No. 13. SPHENE (+ve) $CaTiSiO_5$

Also known as *titanite*.
Some substitution of Sr, Ba, Th or the rare earths for Ca, Fe^{3+} and Al for Ti, and OH and F for O is possible.

Monoclinic, $\beta = 120°$. Distinct $\{110\}$ cleavages. Euhedral crystals with rhombic cross-sections common. Also anhedral as lensoid or drop-like grains. Twinning (lamellar) with $\{221\}$ composition planes are not uncommon.

Colour in thin section: colourless or pleochroic sugary-brown with $Z > Y > X$.

Colour in detrital grains: sugary or dark brown.

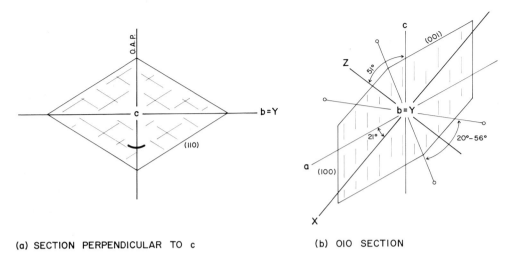

(a) SECTION PERPENDICULAR TO c   (b) OIO SECTION

Fig. 60. Orientation diagrams for sphene.

Optical properties: biaxial +ve. $2V_z = 20°-56°$.
$\alpha = 1.840-1.950$, $\beta = 1.870-2.034$, $\gamma = 1.943-2.110$.
$\delta = 0.100-0.192$. $X \wedge a = $ ca. $21°$, $Y = b$, $Z \wedge c = $ ca. $51°$.

Orientation diagrams. *Section perpendicular to c* (Fig. 60a): off-centred figure which may be close to an optic-axis figure; δ' usually high; symmetrical extinction with respect to two cleavages; traces of twin-lamellae may be visible parallel to cleavages.
(010) section (Fig. 60b): flash figure; δ (very high) $= 0.1-0.192$; a cleavage trace may be visible with $Z \wedge c = 51°$; twin-lamellae may be visible oblique to cleavage trace.
(100) sections display no cleavage.
Optic-axis figures obtained in sections slightly oblique to (100) and approximately parallel to (001).

Occurrence. Very common as an accessory mineral in intermediate and acid plutonic igneous rocks, particularly syenites. Also in a wide range of metamorphic rocks, and as detrital grains.

Distinguishing features. The very high relief and δ and crystal shapes are distinctive. Superficially similar to the carbonates in thin section, but carbonates have a lower R.I. and often "twinkle". May be confused with monazite which has a lower δ, and with detrital grains of xenotime, cassiterite or rutile, all of which are uniaxial.

No. 14. ZIRCON (+ve) $ZrSiO_4$

Usually contains Hf, and often Y, Th, U, and Fe^{3+}.

Tetragonal. Poor $\{110\}$ cleavages. Usually as small euhedral or subhedral prismatic crystals with $\{110\}$ dominating.

Colour in thin section: colourless or pale brown.

Colour in detrital grains: colourless, yellow, brown, pink, or purple; pleochroism $\epsilon > \omega$.

Optical properties: uniaxial +ve.
$\omega = 1.92–1.96$, $\epsilon = 1.96–2.02$.
$\delta = 0.04–0.06$.

Orientation diagrams. *Section parallel to c, and elongate detrital grains* (Fig. 61): flash figure; δ (high) = 0.04–0.06; straight extinction, slow along.
Uniaxial-cross figures obtained in sections cut ⊥ c, with equidimensional cross-sections (often square) and low interference colours. Usually not easy to obtain in thin sections because of small size of crystals, or in detrital grains because of grain shape.

Occurrence. A common accessory mineral in plutonic igneous rocks, particularly granites and syenites. May form large crystals in pegmatites. Also as an accessory mineral in schists and gneisses. It is very resistant to weathering, and is abundant as detrital grains in many sediments, and may be preserved through several cycles of deposition. Detrital grains, sometimes with overgrowths, may also be recognised in metamorphic rocks and granites. Zircon produces pleochroic haloes in biotite. Radioactive elements in zircon may destroy the atomic structure producing "metamict" isotropic pseudomorphs.

Distinguishing features. Extreme relief, high δ, crystal shapes and straight extinction are distinctive. May be confused with cassiterite and rutile, but these minerals are usually deeper brown, and have a higher δ. Xenotime has a lower R.I. and higher δ.

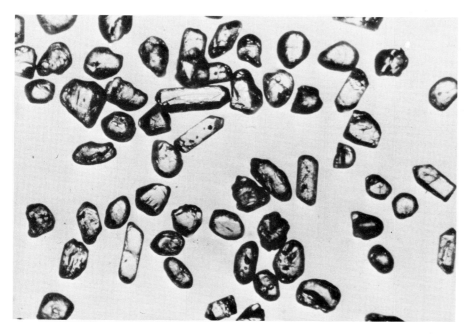

Fig. 61. Detrital grains of zircon. Note the prisms with pyramidal terminations **well** preserved. Pleistocene sands, Westport, New Zealand (view measures 1.7 mm × 1.3 mm). Plane-polarised light.

No. 15. SAPPHIRINE (−ve or +ve) $(Mg,Fe)_2 Al_4 O_6 [SiO_4]$

Monoclinic, $\beta = 126°$. No good cleavage. Anhedral, or tablets parallel to (010). Lamellar twins with $\{010\}$ composition planes not common.

Colour in thin section: weakly pleochroic from colourless, yellow or pink to blue or blue-green usually with $Z > Y > X$.

Optical properties: biaxial −ve (more rarely +ve). $2V_x = 48°−114°$.
$\alpha = 1.701−1.729$, $\beta = 1.703−1.732$, $\gamma = 1.705−1.734$.
$\delta = 0.004−0.010$. $X \wedge a = 42°−45°$, $Y = b$, $Z \wedge c = 6°−9°$.

Occurrence. A rare mineral usually found in high-grade regionally and thermally metamorphosed Al-rich rocks, and often associated with spinel, corundum, and cordierite.

Distinguishing features. High relief, low δ, blue colour and lack of cleavage are distinctive. May be confused with corundum, blue-amphibole, kyanite, tourmaline or dumortierite, but corundum and tourmaline are both uniaxial, blue-amphibole and kyanite have good cleavages, and tourmaline and dumortierite are more strongly pleochroic.

High-temperature calcium silicate minerals

In addition to the more common minerals wollastonite, tremolite, vesuvianite, grossularite, scapolite, epidote, diopside, monticellite, and the humite group, a variety of rare calcium silicate minerals develop at high temperatures as a result of the metamorphism and metasomatism of carbonate rocks. Three of the most common such minerals, spurrite, larnite, and merwinite, are described below briefly.

No. 16. SPURRITE (−ve) $2Ca_2 [SiO_4] \cdot CaCO_3$

Monoclinic, $\beta = 123°$. Distinct $\{001\}$ and imperfect $\{100\}$ cleavages. Usually anhedral, granular. Lamellar twinning with $\{001\}$ composition planes.

Colour in thin section: colourless.

Optical properties: biaxial −ve. $2V_x = 40°$.
$\alpha = 1.640$, $\beta = 1.674$, $\gamma = 1.680$.
$\delta = 0.040$. $X = b$, $Y \wedge c = 33°$, Z approximately parallel to a.

Occurrence. Restricted to high-grade thermally metamorphosed and metasomatised carbonate rocks.

Distinguishing features. Occurrence is distinctive. Distinguished from the associated minerals larnite and merwinite by its −ve character and higher δ.

No. 17. LARNITE (+ve) $Ca_2 [SiO_4]$

Monoclinic, $\beta = 95°$. Distinct $\{100\}$ cleavage. Usually anhedral, granular. Lamellar twinning with $\{100\}$ composition planes.

Colour in thin section: colourless.

Optical properties: biaxial +ve. $2V_z$ moderate.
$\alpha = 1.707$, $\beta = 1.715$, $\gamma = 1.730$.
$\delta = 0.023$. $X \wedge c = 13°$, $Y \wedge a = 8°$, $Z = b$.

Occurrence. Restricted to high-grade thermally metamorphosed and metasomatised carbonate rocks.

Distinguishing features. Occurrence is distinctive. Often associated with spurrite and merwinite. Distinguished from spurrite by its +ve character, and from merwinite by its different cleavage, twinning, extinction angles, and usually higher δ.

No. 18. MERWINITE (+ve) $Ca_3Mg[Si_2O_8]$

Monoclinic, $\beta = 95°$. Perfect $\{010\}$ cleavage. Usually anhedral, granular. Two intersecting sets of lamellar twins with $\{110\}$ composition planes.

Colour in thin section: colourless.

Optical properties: biaxial +ve. $2V_z = 52°-76°$.
$\alpha = 1.702-1.710$, $\beta = 1.710-1.718$, $\gamma = 1.718-1.726$.
$\delta = 0.008-0.023$. $X \wedge c = 36°$, $Y \wedge a = 31°$, $Z = b$.

Occurrence. Restricted to high-grade thermally metamorphosed and metasomatised carbonate rocks.

Distinguishing features. Occurrence is distinctive. Often associated with spurrite and larnite. Distinguished from spurrite by its +ve character, and from larnite by its twinning, cleavage, extinction angles, and usually lower δ.

No. 19. VESUVIANITE (—ve) $Ca_{19}(Mg,Fe,Al)_5Al_8[Si_2O_7]_4[SiO_4]_{10}(OH,F)_{10}$

Also known as *idocrase*.

Tetragonal. Imperfect $\{110\}$ and $\{100\}$ cleavages. Euhedral prismatic to anhedral, granular crystals.

Colour in thin section: colourless, pale brown or green.

Optical properties: uniaxial —ve (rarely +ve or biaxial).
$\omega = 1.705-1.742$, $\epsilon = 1.701-1.736$.
$\delta = 0.001-0.008$.
Anomalous deep-blue, brown, or purple interference colours common.

Occurrence. Typically occurs in thermally metamorphosed or metasomatised carbonate rocks. More rarely occurs in nepheline-syenites, and in veins cutting ultramafic rocks such as serpentinite.

Distinguishing features. High relief, very low δ, anomalous interference colours, poor cleavage, and occurrence distinctive. May be confused with melilite, zoisite or clinozoisite, but zoisite and clinozoisite are +ve, and melilite has a better cleavage and lower R.I.

Vesuvianite that does not display anomalous interference colours is similar to apatite, but has a higher R.I.

The epidote group

The five minerals — zoisite, clinozoisite, epidote, piemontite, and allanite — have in common a structure of independent $[SiO_4]$ and $[Si_2O_7]$ groups which link together chains of AlO_6 and $AlO_4(OH)_2$. Clinozoisite and epidote form a continuous series, and are described together.

No. 20. ZOISITE (+ve) $Ca_2Al_3O[Si_2O_7][SiO_4](OH)$

Fe^{3+} may substitute for up to about 5% of the Al in ferrian or α-zoisite (Myer, 1966). Some Mn substitutes for Al in the variety *thulite*.

Orthorhombic. Perfect $\{100\}$ cleavage (N.B. this is reported as $\{010\}$ in many texts, but X-ray work necessitates the change from the previous a, b, and c to c, a, and b — Deer et al., 1962). Often subhedral aggregates of bladed or fibrous crystals elongate parallel to b.

Colour in thin section: colourless; thulite is pleochroic with X = pink, Y = colourless, Z = yellow.

Optical properties: biaxial +ve. $2V_z = 0°-70°$ (usually $>30°$); a large $2V_x$ has also been reported.
$\alpha = 1.685-1.705$, $\beta = 1.688-1.710$, $\gamma = 1.697-1.725$.
$\delta = 0.004-0.022$ (usually <0.010; the higher values are normally due to a high Fe or Mn content).
$X = b$, $Y = a$, $Z = c$ for Fe-rich ferrian or α-zoisite.
$X = a$, $Y = b$, $Z = c$ for β-zoisite.
Zoisite may display anomalous blue interference colours.

Orientation diagrams. *(001) sections* (Fig. 62a, b): acute bisectrix figure; O.A.P. parallel to perfect cleavage in α form, and perpendicular to cleavage in β form; δ' (very low) < 0.005; may be fast (α-zoisite) or slow (β-zoisite) along, and straight extinction.
(100) sections and cleavage fragments (Fig. 62c, d): obtuse bisectrix (β-zoisite) or flash figure (α-zoisite); δ' is very low to moderate; α-zoisite (100) sections display the highest interference colours; fast along, and straight extinction in euhedral grains.
Optic-axis figures obtained in sections showing good cleavage in α-zoisite, but in sections oblique to cleavage in β-zoisite.

Occurrence. Not as common as clinozoisite—epidote, but widespread, occurring in regionally and thermally metamorphosed (usually impure carbonate) rocks. Zoisite may be a component of saussurite — an alteration product of plagioclase.

Distinguishing features. High relief, low to moderate δ, single cleavage and straight extinction characteristic. Most easily confused with clinozoisite which, however, has inclined extinction. May sometimes be confused with vesuvianite, apatite and melilite, but these minerals are uniaxial and usually —ve.

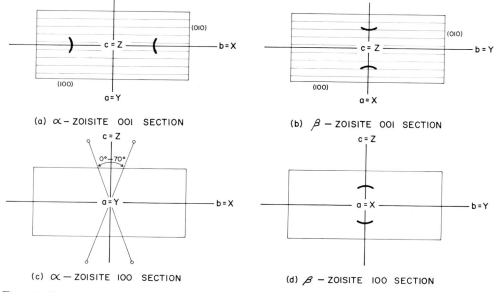

(a) α – ZOISITE 001 SECTION

(b) β – ZOISITE 001 SECTION

(c) α – ZOISITE 100 SECTION

(d) β – ZOISITE 100 SECTION

Fig. 62. Orientation diagrams for zoisite.

No. 21. CLINOZOISITE (+ve) — EPIDOTE (−ve)

$$Ca_2(Al, Fe^{3+})Al_2O[Si_2O_7][SiO_4](OH)$$

There is increasing substitution of Fe for Al from clinozoisite to epidote.

Monoclinic, $\beta = 115°$. Perfect $\{001\}$ and imperfect $\{100\}$ cleavages. Often subhedral bladed or fibrous crystals elongate parallel to b. Also anhedral. Twinning with $\{100\}$ composition plane not common. Zoning common.

Colour in thin section: clinozoisite, colourless; epidote, colourless or yellow-green pleochroic with $Z > Y > X$, or $Y > Z > X$; colour-zoning common.

Colour in detrital grains: clinozoisite, colourless or pale yellow; epidote, yellowish green.

Optical properties: *Clinozoisite*: biaxial +ve. $2V_z = 14°-90°$.
$\alpha = 1.670-1.715$, $\beta = 1.674-1.725$, $\gamma = 1.690-1.734$.
$\delta = 0.005-0.015$. $X \wedge c = 0°-97°$ (in obtuse angle β — decreases with increase in Fe),
$Y = b$, $Z \wedge a = 0°-72°$.
Epidote: biaxial −ve. $2V_x = 64°-90°$.
$\alpha = 1.714-1.751$, $\beta = 1.721-1.784$, $\gamma = 1.730-1.797$.
$\delta = 0.015-0.049$. $X \wedge c = 0°-15°$ (in acute angle β — increases with increase in Fe),
$Y = b$, $Z \wedge a = 25°-40°$.
1st-order white interference colour is absent in both clinozoisite and epidote, and replaced by anomalous blue-grey or yellow.
R.I. and δ increase generally with increase in Fe.

Orientation diagrams. *(010) section* (Fig. 63a): flash figure; δ varies from low for clinozoisite to moderate or high for epidote; inclined extinction $Z \wedge a$ varies from clinozoisite

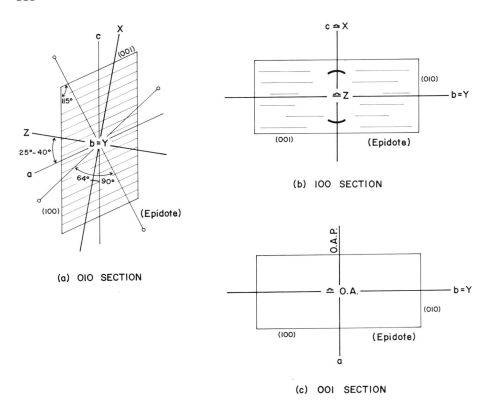

Fig. 63. Orientation diagrams for epidote.

to epidote; a second cleavage parallel to (100) may sometimes be developed.

(100) section (Fig. 63b): obtuse bisectrix figure for epidote, but positions of X and Z vary widely in clinozoisite; O.A.P. across cleavage; δ' varies from low for clinozoisite to moderate or high for epidote; straight extinction; epidote is slow along.

(001) section and cleavage fragments (Fig. 63c): close to optic-axis figure for epidote and some clinozoisite; no cleavage visible; δ' low, often anomalous interference colours; may be slow or fast along the length of euhedral grains.

Occurrence. Common in metamorphic rocks, particularly greenschists and metamorphosed and metasomatised carbonate rocks. Also in veins, as an alteration product in igneous rocks, and in vesicles. May be a component of saussuritised plagioclase. Common as a detrital mineral.

Distinguishing features. High relief and, for epidote, moderate to high δ, single cleavage, and inclined extinction in (010) sections are characteristic. The characteristic colour of epidote, and the lack of 1st-order white interference colour (visible even at the edge of highly birefringent grains) are very distinctive. Clinozoisite is distinguished from epidote by its lower δ and +ve character. Olivine, pyroxene, amphibole and zoisite may be confused with clinozoisite—epidote. It is distinguished from pyroxene and amphibole by its single cleavage with the O.A.P. across it, from olivine and zoisite by its inclined extinction in some sections, and from olivine, pyroxene, and amphibole by its anomalous interference colours and characteristic yellow-green colour (epidote).

No. 22. PIEMONTITE (+ve) $Ca_2(Mn,Fe^{3+},Al)_2AlO[Si_2O_7][SiO_4](OH)$

Monoclinic, $\beta = 115°$. Perfect $\{001\}$ cleavage. Subhedral crystals elongate parallel to b.

Colour in thin section: pink or purple, pleochroic with $Z > Y > X$, or $Y > X > Z$.

Optical properties: biaxial +ve. $2V_z = 64°-85°$.
$\alpha = 1.732-1.794$, $\beta = 1.750-1.807$, $\gamma = 1.762-1.829$.
$\delta = 0.025-0.088$. $X \wedge c = 2°-7°$ (in acute angle β), $Y = b$, $Z \wedge a = 28°-33°$.

Orientation diagrams. Similar to epidote, but +ve.

Occurrence. In low-grade schists, altered igneous rocks, and metamorphic and hydro-thermal manganese deposits.

Distinguishing features. Similar to clinozoisite—epidote, but its colour and pleochroism are distinctive.

No. 23. ALLANITE (−ve, rarely +ve) $(Ca,Ce)_2(Fe^{2+},Fe^{3+})Al_2O[Si_2O_7][SiO_4](OH)$

Also known as *orthite*.

Monoclinic, $\beta = 116°$. Imperfect $\{001\}$ and $\{100\}$ cleavages. Subhedral crystals elongate parallel to b; also anhedral. Zoning common.

Colour in thin section: dark brown or red-brown, pleochroic usually with $Z > Y > X$, or $Y > Z > X$; colour-zoning common.

Optical properties: biaxial −ve (rarely +ve). $2V_z = 40°-123°$.
$\alpha = 1.690-1.813$, $\beta = 1.700-1.857$, $\gamma = 1.706-1.891$.
$\delta = 0.013-0.078$. $X \wedge c = 1°-42°$, $Y = b$, $Z \wedge a = 27°-68°$

Orientation diagrams. Similar to epidote, but lacks a well-developed cleavage.

Occurrence. An accessory mineral in granites, diorites, alkaline igneous rocks, carbona-tites, and pegmatites. Also in metamorphosed carbonate rocks. Radioactive elements in allanite may destroy the atomic structure, reducing the mineral to an isotropic "meta-mict" state. Allanite inclusions in biotite produce pleochroic haloes.

Distinguishing features. The high or very high relief, moderate to very high δ, and brown colour are distinctive. Orientation diagrams, and lack of good cleavage distinguish allanite from brown amphibole.

No. 24. LAWSONITE (+ve) $CaAl_2(OH)_2[Si_2O_7]H_2O$

The structure of lawsonite is similar to that of the epidote group, and contains chains of Al surrounded by O and OH. As for epidote, the direction of the chains is designated b, but in some other texts this direction is designated c.

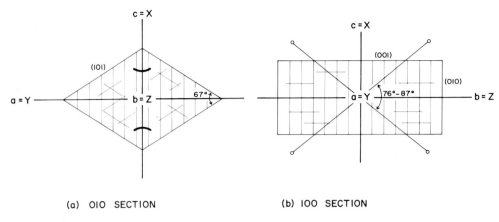

(a) OIO SECTION   (b) IOO SECTION

Fig. 64. Orientation diagrams for lawsonite.

Orthorhombic. Distinct {010} and {100}, and poor {101} cleavages. Euhedral crystals may be prismatic and elongate parallel to b, or (010) plates. Intersecting sets of twins with {101} composition planes common.

Colour in thin section: usually colourless, but may be pleochroic with X = blue, Y = yellow, and Z = colourless.

Optical properties: biaxial +ve. $2V_z = 76° - 87°$.
$\alpha = 1.665$, $\beta = 1.674$, $\gamma = 1.685$.
$\delta = 0.020$. $X = c$, $Y = a$, $Z = b$.

Orientation diagrams. *(010) section* (Fig. 64a): acute bisectrix figure; δ' (low) = 0.009; good {100} and poor {101} cleavages visible; fast direction parallel to good cleavage.
(100) section (Fig. 64b): flash figure; δ (moderate) = 0.020; sections may be elongate parallel to b, or parallel to the {010} cleavage; trace of poor {101} cleavage may be visible; fast parallel to good cleavage; straight extinction.
(001) section: extinction parallel to two good cleavages at right angles to each other; obtuse bisectrix figure.

Occurrence. Lawsonite is a comparatively rare mineral, but it is important in petrology and diagnostic of the low-temperature high-pressure lawsonite—albite facies of regional metamorphism. It is often found in glaucophane schists; also frequently present in metamorphic rocks that have retained their pre-metamorphic fabric and texture.

Distinguishing features. Moderate to high relief, moderate δ, cleavages, and occurrence are characteristic. May be confused with zoisite and prehnite, but zoisite has anomalous interference colours, and prehnite has a higher δ and lower R.I.

No. 25. PUMPELLYITE (+ve, also −ve)
$$Ca_4(Mg,Fe^{2+})(Al,Fe^{3+})_5[Si_2O_7]_2[SiO_4]_2O(OH)_3 2H_2O$$

Monoclinic, $\beta = 97°$. Distinct {001} and imperfect {100} cleavages. Crystals elongate parallel to b. Twinning with {001} and {100} composition planes.

Colour in thin section: pleochroic, colourless to green or less commonly brown, with $Y > Z > X$; may be colour-zoned.

Optical properties: biaxial +ve (less commonly —ve). $2V_z = 10°—92°$; brown Fe-rich varieties may have $2V_x = 0°—80°$.
$\alpha = 1.665—1.728$, $\beta = 1.670—1.748$, $\gamma = 1.683—1.754$.
$\delta = 0.010—0.028$. $X \wedge a = 5°—37°$, $Y = b$, $Z \wedge c = 0°—30°$. Brown Fe-rich varieties may also have the O.A.P. $\perp$ (010).
R.I., $2V_z$, and extinction angles increase with Fe; $\beta > 1.73$ for —ve varieties.

Orientation diagrams. *(010) section, +ve varieties* (Fig. 65): flash figure; δ (low to moderate) = $0.010—0.020$; inclined extinction with fast onto the better cleavage $5°—37°$.
Sections parallel to b: straight extinction; the trace of a cleavage parallel to b may be visible; slow or fast along.

Occurrence. Usually as small crystals in a wide range of low-grade metamorphic rocks that in many cases have retained their original fabric and texture. Also occurs as a secondary mineral in volcanic rocks and as an alteration product of biotite.

Distinguishing features. May be confused with the epidote-group minerals. Colourless (Fe-poor) pumpellyite is distinguished from zoisite and clinozoisite by its lower R.I., and from zoisite by its inclined extinction in (010) sections. Green pumpellyite has a lower R.I. than epidote and is usually +ve. The angle between the $\{001\}$ and $\{100\}$ cleavages is $83°$ for pumpellyite and $65°$ for epidote.

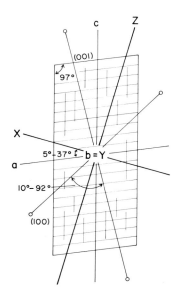

OIO SECTION (commoner varieties)

Fig. 65. Orientation diagram for pumpellyite.

No. 26. MELILITE (−ve and +ve)

A series from *gehlenite* $Ca_2Al_2SiO_7$ to *åkermanite* $Ca_2MgSi_2O_7$. Na may substitute for Ca, and Fe for Al.

Tetragonal. Imperfect {001} cleavage. Euhedral—subhedral tabular crystals parallel to (001) with square cross-sections common.

Colour in thin section: usually colourless; may be yellow, pleochroic with $\epsilon > \omega$.

Optical properties: uniaxial −ve or +ve.
In the gehlenite—åkermanite series, ω = 1.632—1.669, ϵ = 1.640—1.658. δ = 0—0.011, gehlenite being −ve, akermanite +ve. The change from +ve to −ve takes place at $Ge_{48}Ak_{52}$ which mineral is isotropic. Na affects these properties so that the R.I. may be as low as ϵ = 1.616; Fe increases the R.I. which may be as high as ω = 1.7.
Interference colours may be anomalous blue.
Straight extinction and either (−ve) slow along, or (+ve) fast along the cleavage.

Occurrence. Rather rare, but may be abundant in some undersaturated lavas. Also found in thermally metamorphosed and metasomatised carbonate rocks.

Distinguishing features. Tabular crystals, uniaxial character, moderate relief, anomalous interference colours, and occurrence are distinctive. May be confused with vesuvianite or zoisite, but both these minerals have a higher R.I., and zoisite is biaxial.

B. CYCLOSILICATES

In the cyclosilicates, the $[SiO_4]$ tetrahedral groups are linked together to form rings. A hexagonal ring $[Si_6O_{18}]$ is the dominant structural feature of the minerals beryl, cordierite, and tourmaline, whereas a tetragonal ring $[Si_4O_{12}]$ is the basis of the structure of axinite.

No. 27. BERYL (−ve) $Be_3Al_2[Si_6O_{18}]$

The precious variety *emerald* contains Cr.

Hexagonal. Imperfect {0001} cleavage. Usually six-sided prismatic euhedral crystals.

Colour in thin section: colourless.

Colour in detrital grains: usually blue-green, pleochroic with $\omega > \epsilon$.

Optical properties: uniaxial −ve.
ω = 1.568—1.608, ϵ = 1.564—1.600.
δ = 0.004—0.009.
Prismatic crystals have straight extinction and are length fast.

Occurrence. Principally in granite pegmatites, but also in metasomatised schists and carbonate rocks.

Fig. 66. Sector twinning in cordierite from a dacite. Reproduced with permission from Zeck (1972). Crossed-polarised light.

Distinguishing features. Low to moderate relief, uniaxial −ve character, and low δ distinctive. May be confused with quartz or apatite, but quartz has a lower R.I. and is +ve, and apatite has a higher R.I.

No. 28. CORDIERITE (−ve and +ve) $(Mg,Fe)_2 Al_3 [Si_5 AlO_{18}]$

Orthorhombic. Imperfect $\{010\}$ cleavage. Anhedral or euhedral short prismatic pseudo-hexagonal crystals made up of sector twins (Fig. 66). Twinning common; either sector twins or multiple twinning (often intersecting sets) with composition planes approximately parallel to $\{110\}$ or $\{130\}$. The sector twins may radiate from a central point at intervals of 30°, 60° or 120° (Figs. 66 and 67), and have been interpreted by Zeck (1972) to be the result of transformation from a high-temperature hexagonal state.

Colour in thin section: usually colourless; may have yellow pleochroic haloes around inclusions of zircon and apatite.

Colour in detrital grains: weakly pleochroic with X = colourless or pale yellow, Y and Z = violet or blue.

Optical properties: biaxial −ve or +ve. $2V_x = 42° − 104°$.
$\alpha = 1.522 − 1.560, \beta = 1.524 − 1.574, \gamma = 1.527 − 1.578.$
$\delta = 0.005 − 0.018.$ $X = c, Y = a, Z = b.$

Orientation diagrams. *(001) section* (Fig. 67): bisectrix figure; δ' (very low) = 0.003; may show sector twins or intersecting sets of lamellar twins; orientations shown in Fig. 67a, b are for twins with composition planes ca. $\{110\}$ − interchange Y and Z positions for ca. $\{130\}$ composition planes.
Prismatic sections may display simple or multiple twin-lamellae parallel to the c-axis.

Occurrence. Most commonly found in thermally metamorphosed pelitic sediments, often as ovoid poikiloblastic grains. Also in pelitic rocks regionally metamorphosed under high geothermal gradients, and in metasomatised(?) gneisses with anthophyllite. Present in some igneous rocks, especially contaminated gabbros, and in some granite pegmatites. Alters to pinite, a mixture of sericite and chlorite.

Distinguishing features. The sector twinning and yellow pleochroic haloes are very distinctive, but not always present. In their absence, cordierite may be confused with quartz and plagioclase feldspar. Quartz is uniaxial, but the distinction from plagioclase may be very difficult, especially if the multiple lamellar twinning is present. Lack of cleavage suggests cordierite, but it may be necessary to observe the pleochroism of cordierite in thick sections to distinguish the two. A staining method to distinguish cordierite from quartz

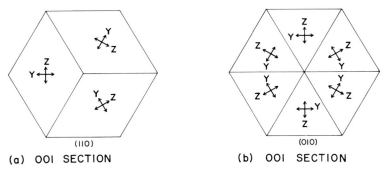

(a) OOI SECTION   (b) OOI SECTION

Fig. 67. Orientation diagrams for cordierite with sector twinning; ca. $\{110\}$ composition planes.

and plagioclase has been described by Gregnanin and Viterbo (1965); the thin section is etched with HF vapour and stained with a solution of HCl and potassium ferricyanide which turns cordierite azure but leaves quartz and plagioclase colourless.

No. 29. TOURMALINE (—ve) $Na(Mg,Fe,Mn,Li,Al)_3 Al_6 [Si_6 O_{18}](BO_3)_3(OH,F)_4$

Common Fe-tourmaline is known as *schorl*, Mg-tourmaline as *dravite*, and alkali-tourmaline as *elbaite*.

Trigonal. No good cleavage. Usually elongate prismatic crystals with a hexagonal or triangular (convexly curved sides) cross-section.

Colour in thin section: variable, and strongly pleochroic with $\omega > \epsilon$; schorl is black, brown, green, or blue; dravite is brown, yellow or colourless; elbaite is colourless or pink. Often colour-zoned.

Colour in detrital grains: variable as above, but deeper colours.

Optical properties: uniaxial —ve.
$\omega = 1.635—1.675; \epsilon = 1.610—1.650.$
$\delta = 0.017—0.035.$
R.I. and δ increase with Fe content. An unusual Fe^{3+}-rich tourmaline with $\omega = 1.735$ and $\epsilon = 1.655$, and an authigenic tourmaline with $\omega = 1.633$ and $\epsilon = 1.621$ have been reported.
Prismatic sections have straight extinction, and are length fast.

Occurrence. Schorl is an accessory in many granites, pegmatites, schists, and gneisses, and is an important constituent of some metasomatic rocks. It is also a common detrital mineral. Dravite is found in metamorphosed and metasomatised carbonate rocks. Elbaite is found in granite-pegmatites in association with other Li minerals such as lepidolite.

Distinguishing features. *Schorl* — the moderate relief and δ, strong colour and pleochroism, lack of cleavage, and crystal shape are very distinctive. It may be confused with (001) sections of biotite or (100) sections of hornblende. Both these minerals differ from tourmaline in other orientations in which they display good cleavages, and a maximum absorption parallel to their length; hornblende is biaxial. *Dravite* — may be confused with chondrodite with which it may be associated. Dravite is distinguished by its uniaxial character. *Elbaite* — the moderate δ distinguishes elbaite from other colourless uniaxial minerals of similar relief.

No. 30. AXINITE (—ve) $(Ca,Fe,Mg,Mn)_3 Al_2 BO_3 [Si_4 O_{12}]OH$

Triclinic, $\alpha = 88°, \beta = 81°, \gamma = 78°$. Distinct $\{100\}$ and a number of other very poor cleavages. Usually euhedral wedge-shaped crystals with $\{010\}$ and $\{011\}$ well developed.

Colour in thin section: colourless, or yellow and violet pleochroic with $Y > X > Z$.

Optical properties: biaxial —ve. $2V_x = 63°—80°$.
$\alpha = 1.659—1.693, \beta = 1.665—1.701, \gamma = 1.668—1.704.$
$\delta = 0.007—0.014$. X is approximately normal to $(\bar{1}11)$, but this position varies with composition.
Inclined extinction to crystal outlines and cleavage in most sections.

Occurrence. Most common in thermally metamorphosed and metasomatised carbonate rocks, and in metasomatised igneous rocks. Also found in quartz veins.

Distinguishing features. The crystal shape, high relief, low δ, large $-$ve $2V$, and inclined extinction are distinctive.

C. PHYLLOSILICATES

In the phyllosilicates, each $[SiO_4]$ tetrahedral group shares three of its oxygen atoms with three other tetrahedra to form a continuous sheet of general formula $[Si_2O_5]$. Phyllosilicates described here are: the micas; the chlorites and serpentines; pyrophyllite; talc; stilpnomelane; the brittle-micas margarite and clintonite; prehnite; apophyllite; and the clay minerals.

The micas

The micas are characterised by a composite structure of two $[Si_2O_5]$ sheets between which are sandwiched Fe, Mg, Al, or Li cations. K or Na atoms lie between, and (OH) ions within the composite sheets. The sheets may be stacked upon one another in a number of ways resulting in a variety of polymorphs, but a study of these is beyond the scope of this book. The

Fig. 68. Mottled extinction in biotite, Constant Gneiss, New Zealand (view measures 3.3 mm × 2.1 mm). Crossed-polarised light.

"brittle-micas" (margarite and clintonite) have a similar structure with Ca atoms lying between the composite sheets.

The micas described below are muscovite (including sericite, phengite and fuchsite), paragonite, the Li-mica lepidolite, biotite—phlogopite, and glauconite. Characteristics common to them all include: a single perfect cleavage parallel to the sheet structure; X approximately perpendicular to (001) — hence all micas have a slow direction parallel to the cleavage traces in thin section; biaxial —ve, and sometimes pseudo-uniaxial — hence interference colours are always lower in basal sections than side sections; mottled extinction (Fig. 68) caused by the microscopic buckling of cleavage planes during thin-section making. However, these characteristics, even in combination, *do not* distinguish micas from all other minerals.

No. 31. MUSCOVITE (—ve) $K_2Al_4[Si_6Al_2O_{20}](OH,F)_4$

Varieties include *phengite* (Si replaces Al in the sheet structure and Mg or Fe replaces some of the Al sandwiched between the composite sheets), *fuchsite* (Cr-rich mica), and *sericite* (fine-grained muscovite).

Monoclinic, β usually = 95°. Perfect {001} cleavage. Often subhedral tabular crystals with {001} well developed; sometimes euhedral tablets with {001}, {110}, and {010} forms.

Colour in thin section: usually colourless, but fuchsite may be pale to dark emerald green with $X < Y \leqslant Z$.

Colour in detrital grains: usually colourless.

Optical properties: biaxial – ve. $2V_x = 30° - 47°$ (but as low as 0° in some phengite). $\alpha = 1.552 - 1.578$, $\beta = 1.582 - 1.615$, $\gamma = 1.587 - 1.617$. $\delta = 0.036 - 0.049$. $X \wedge c = 2° - 4°$, $Y \wedge a = 1° - 3°$, $Z = b$. R.I. and δ increase with Fe and Cr, and $2V$ decreases with Mg and Fe. Hence phengite and fuchsite are high R.I. muscovites.

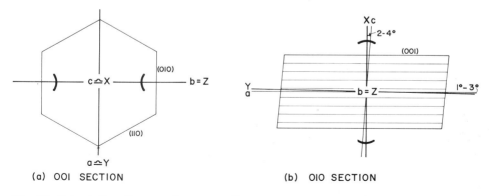

(a) 001 SECTION   (b) 010 SECTION

Fig. 69. Orientation diagrams for muscovite.

Orientation diagrams. *(001) section and cleavage fragments* (Fig. 69a): acute bisectrix figure; δ' (low) up to 0.006; extinction positions often indistinct due to proximity of optic axes; extinction may be wavy due to bending of cleavages; cleavage *not* visible.
(010) section (Fig. 69b): obtuse bisectrix (flash) figure; δ' (high) up to 0.037; good cleavage, slow along and almost straight extinction (mottled); if α has a low value, this section "twinkles" on rotation of stage.
(100) section: since Y ca. = Z, this section is almost identical to the (010) section; δ up to 0.049.
Optic-axis figures obtained in sections slightly oblique to (001), and in which cleavage is not visible.

Occurrence. Widespread in regionally metamorphosed rocks, especially those derived from pelitic sediments. At high grades, muscovite may break down to form sillimanite and K-feldspar; sillimanite frequently occurs as swarms of minute fibres embedded in muscovite. Phengite is common in high-pressure low-temperature metamorphic rocks. Muscovite is less common in thermally metamorphosed rocks. It is a common constituent of granites, pegmatites, and greisens, but is unstable in the volcanic environment of high temperatures and low pressures. It is resistant to weathering, and is therefore a common detrital mineral. It may also form during diagenesis. Many minerals (especially feldspar, nepheline, andalusite, kyanite and cordierite) alter to fine-grained muscovite (sericite). Some sericite may, however, be paragonite or illite.

Notes. Sections of muscovite displaying cleavage and high interference colours are easily identified, but basal sections are often overlooked or misidentified. Loose grains of muscovite always lie on their cleavage, do not display high interference colours, and do not allow measurement of α. Rocks cut normal to a schistosity defined by muscovite will seldom display basal sections.

Distinguishing features (see Table II). Perfect single cleavage, straight extinction (mottled), slow along, high δ, and lack of colour are characteristic. Fuchsite is usually prominent as a green mineral in hand specimen. Muscovite is easily confused with phlogopite, talc, paragonite, the Li-micas, and pyrophyllite. Phlogopite and talc have a small $2V$, but may still be confused with some phengites. Talc and phlogopite have distinctive occurrences, and talc is soft and has a soapy feel in hand specimen. Pyrophyllite has a higher $2V$. Lepidolite may have a lower R.I. than muscovite, is often pink or purple in hand specimen, and occurs only in pegmatites and associated rocks. Paragonite is optically indistinguishable from muscovite. Chemical, X-ray, or percussion tests may be necessary to distinguish these minerals There is a stain test to distinguish muscovite from paragonite (Laduron, 1971).

No. 32. PARAGONITE (—ve) $Na_2Al_4[Si_6Al_2O_{20}](OH)_4$

Monoclinic, β = 95°. Perfect {001} cleavage. Usually fine-grained aggregates of crystals with {001} dominating.

Colour in thin section: colourless.

Optical properties: biaxial —ve. $2V_x = 0°-46°$.
α = 1.564—1.580, β = 1.594—1.609, γ = 1.600—1.609.
δ = 0.028—0.038. $X \wedge c$ = ca. 5°, Y = ca. a, Z = b.

Orientation diagrams. Same as for muscovite (No. 31).

Occurrence. Occurs in schists and sediments. Some "sericite" may be paragonite. Because of the difficulty of distinguishing this mineral from muscovite, it may be more widespread than generally believed.

TABLE II

Summary of distinctions between some of the phyllosilicates

A. With mottled extinction and high birefringence

Muscovite (31)	colourless; usually larger $2V$ than talc and phlogopite; smaller $2V$ than pyrophyllite; difficult to distinguish from paragonite and lepidolite; *var.* fuchsite is green in hand specimen
Paragonite (32)	cannot be distinguished optically from muscovite
Lepidolite (33)	similar to muscovite, but usually pink or violet in hand specimen, and restricted in occurrence
Phlogopite and biotite (34)	brown or green pleochroic except for some colourless phlogopite which has a restricted occurrence; $2V$ very small; much higher δ than chlorite
Talc (39)	colourless; usually smaller $2V$ than muscovite; distinctive occurrence and soapy feel in hand specimen
Pyrophyllite (38)	colourless; distinguished from micas by its larger $2V$

B. With less pronounced mottled extinction and low birefringence

Chlorite (36)	green (or yellow) pleochroic; often anomalous interference colours
Serpentine (37)	not so green or pleochroic as chlorite; always normal interference colours; some serpentine is fibrous

C. Without mottled extinction

Stilpnomelane (40)	similar to biotite	
Margarite (41) and clintonite (42)	δ moderate, between that of the normal micas and the chlorites	less perfect $\{001\}$ cleavage than in the other phyllosilicates
Prehnite (43)	similar to muscovite, but fast along and $+ve$	

Distinguishing features (see Table II). Same as for muscovite. Cannot be distinguished optically from muscovite, and chemical, X-ray, or staining methods (Laduron, 1971) must be used.

No. 33. LEPIDOLITE (—ve) $K_2(Li,Al)_{5-6}[Si_{6-7}Al_{2-1}O_{20}](OH,F)_4$

Fe may be present in small quantities.

Monoclinic, β = 100°. Perfect {001} cleavage. As large tabular crystals or fine-grained aggregates with {001} dominating.

Colour in thin section: colourless.

Colour in detrital grains: may be pink or lilac.

Optical properties: biaxial —ve. $2V_x$ = 0°—58° (usually 30°—50°).
α = 1.525—1.548, β = 1.548—1.585, γ = 1.551—1.587.
δ = 0.018—0.039. X ca. = c, Y = b, Z ca. = a.
R.I. increases with Fe content.

Orientation diagrams. *(001) section and cleavage fragments*: similar to muscovite, but orientation of O.A.P. is different (but same as for biotite — Fig. 70a); orientation of O.A.P. difficult to determine except in euhedral tablets or by percussion tests.
(010) and (100) sections: similar appearance to muscovite, but generally a slightly lower δ (up to 0.039).

Occurrence. Mainly in granite-pegmatites, aplites, and associated rocks.

Distinguishing features (see Table II). Very similar to muscovite, and chemical or X-ray tests may be necessary to distinguish them. Lepidolite is, however, distinguished by its pink or purple colour in hand specimen, and its restricted occurrence. It usually has a lower R.I. and δ than muscovite.

No. 34. BIOTITE (—ve)
$$K_2(Fe^{2+},Mg)_6(Fe^{3+},Al,Ti)_{0-2}[Si_{6-5}Al_{2-3}O_{20}]O_{0-2}(OH,F)_{4-2}$$
and PHLOGOPITE (—ve) $K_2(Mg,Fe^{2+})[Si_6Al_2O_{20}](OH)_4$

Phlogopite forms a series with biotite. Phlogopite has Mg/Fe > 2/1.

Monoclinic, β = 100°. Perfect {001} cleavage. Often subhedral tabular with {001} dominating; euhedral crystals are tablets with {001}, {110}, and {010} forms. Twinning with {001} composition plane – twin plane is {110} .

Colour in thin section: biotite is strongly pleochroic, pale to dark brown or green with X < Y ⩽ Z. Phlogopite is colourless or pale brown or green. Reddish-brown varieties usually have a high Ti and Fe content, and green ones usually have a high Fe^{3+} content. Pleochroic haloes are common around inclusions of radioactive minerals such as zircon.

Colour in detrital grains: as above but deeper colours.

Optical properties: biaxial —ve. $2V_x$ = 0°—25° (usually <10°).
α = 1.530—1.625, β = 1.557—1.696, γ = 1.558—1.696.

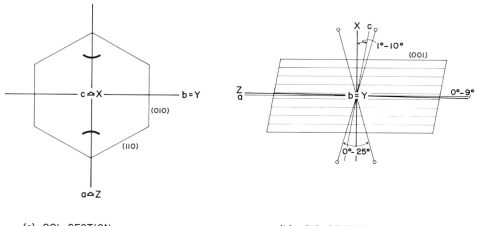

(a) OOI SECTION   (b) OIO SECTION

Fig. 70. Orientation diagrams for biotite—phlogopite.

δ = 0.028—0.078. $X \wedge c$ = $1°$—$10°$, $Y = b$, $Z \wedge a$ = $0°$—$9°$. Occasionally $Z = b$, and Y ca. = a.

R.I. and δ increase generally with Ti and Fe, especially Fe^{3+}.

Orientation diagrams. *(001) section and cleavage fragments* (Fig. 70a): acute bisectrix figure; δ' (very low) usually 0.001; since Y ca. = Z, pleochroism is weak or absent; cleavage not visible; extinction may be wavy due to bending of cleavages.

(010) section (Fig. 70b): flash figure; δ (high or very high) up to 0.078; good cleavage, slow along, and almost straight extinction (mottled); strongly pleochroic with $Z > X$ in coloured members; may display twin-lamellae parallel to cleavage in members whose extinction angle is measurable.

(100) section: since Y ca. = Z, this section is identical to the (010) section, but always has straight extinction.

Optic-axis figures obtained on (001) sections in which cleavage is not visible.

Occurrence. *Biotite* — present in a very wide range of plutonic igneous rocks, but particularly common in granites. Also as phenocrysts (often showing signs of corrosion) in volcanic rocks. Fe content of biotite is generally higher in the more felsic igneous rocks. Very widespread in metamorphic rocks, especially metamorphosed pelites; forms at the onset of thermal metamorphism, but slightly later with the higher pressures of regional metamorphism. At high grades, it may break down to sillimanite, the sillimanite often being embedded in the biotite. It alters to chlorite, but nevertheless is common as a detrital mineral in sediments. *Phlogopite* — more restricted in occurrence. Most common in metamorphosed impure carbonate rocks, in ultrabasic rocks — especially kimberlite, and in some volcanics.

Notes. Sections of biotite and phlogopite displaying cleavage, pleochroism, and high interference colours, are easily identified, but basal sections are often overlooked or misidentified. Loose grains always lie on their cleavage, appear almost isotropic, and do not allow measurement of α. Rocks cut normal to a schistosity defined by biotite will seldom display basal sections.

Distinguishing features (see Table II). Perfect single cleavage, almost straight extinction (mottled), length-slow nature, high δ, and pleochroism are characteristic. Biotite is not easily confused with other minerals, but basal sections may be confused with tourmaline (lacks cleavage) and hornblende (the confusing (100) sections of hornblende provide good figures indicating a moderate to large $2V$). Stilpnomelane is very similar to biotite, but has a less perfect $\{001\}$ cleavage, does not display mottled extinction, and is "brittle" in hand specimen. Phlogopite may be confused with muscovite, talc, paragonite, pyrophyllite and lepidolite. However, except for talc, phlogopite is distinguished by its usually smaller $2V$ and occurrence; talc is soft and soapy to the touch in hand specimen.

No. 35. GLAUCONITE (—ve)

$$(K,Na,Ca)_{1.2-2.0}(Fe^{3+},Al,Fe^{2+},Mg)_4[Si_7AlO_{20}](OH)_4 \cdot n(H_2O)$$

Monoclinic, $\beta = 100°$. Perfect $\{001\}$ cleavage. As fine-grained aggregates in pellets.

Colour in thin section: green, pleochroic with $X < Y = Z$.

Optical properties: biaxial —ve. $2V_x = 0°-20°$.
$\alpha = 1.585-1.616$, β ca. $= \hat{\gamma} = 1.600-1.644$.
$\delta = 0.014-0.032$. $X \wedge c$ up to $10°$, $Y = b$, Z ca. $= a$.

Orientation diagrams. Usually too fine grained to determine properties precisely.

Occurrence. Restricted to sediments. Forms as a diagenetic mineral under marine conditions, often as sand-sized pellets. A major constituent of greensands.

Distinguishing features. Its green colour, fine-grained nature and occurrence are distinctive. Distinguished from chlorite by its higher δ.

No. 36. CHLORITE (+ve or —ve) $(Mg,Fe,Al)_{12}[(Si,Al)_8O_{20}](OH)_{16}$

Mn or Cr may also be present

Usually monoclinic with $\beta = 97°$. Perfect $\{001\}$ cleavage. Euhedral—subhedral pseudo-hexagonal tabular crystals, scaly aggregates, or radiating aggregates, with $\{001\}$ dominating. Twinning with $\{001\}$ composition planes.

Colour in thin section: generally colourless, pale to dark green, pleochroic with $X = Y > Z$ (+ve varieties), or $Z = Y > X$ (—ve varieties); Mn-rich chlorites are usually orange, and Cr-rich chlorites, pink or violet in colour.

Colour in detrital grains: as above, but deeper colours in coarse material.

Optical properties: biaxial +ve or —ve. $2V_z = 0°-45°$ or $2V_x = 0°-30°$.
$\alpha = 1.562-1.671$, $\beta = 1.562-1.690$, $\gamma = 1.565-1.690$.
$\delta = 0.00-0.015$. $X \wedge a$ (+ve) or $Z \wedge a$ (—ve) $<3°$; $Y = b$, $Z \wedge \perp (001)$ (+ve) or $X \wedge \perp (001)$ (—ve) $< 3°$.

Terminology. The compositional and structural variation in the chlorites is wide-ranging, and numerous varietal names have been adopted, e.g. *pennine, chlinochlore,* and *prochlorite* or *ripidolite*. To classify the chlorites and use varietal names properly requires both chemical and X-ray work (Phillips, 1964). However, Albee (1962) has shown that the

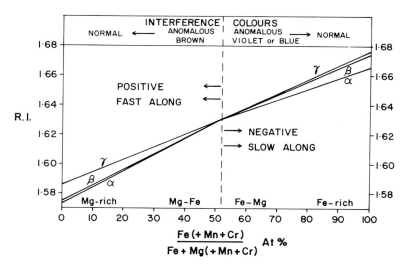

Fig. 71. Correlation of optical properties and chemistry in chlorite. Redrawn with permission from Albee (1962, fig. 4).

Fe/Fe+Mg ratio of most chlorites can be determined optically on the basis of R.I., sign, and interference colour (Fig. 71); his subdivision into Mg-rich, Mg—Fe, Fe—Mg, and Fe-rich chlorites is adopted here:

Mg-rich chlorite: R.I. ca. 1.562—1.60; +ve 2*V*, fast along and normal interference colours.

Mg—Fe chlorite: R.I. ca. 1.60—1.63; +ve 2*V*, fast along, and anomalous brown interference colours.

Fe—Mg chlorite: R.I. ca. 1.63—1.65; —ve 2*V*, slow along, and anomalous violet or blue interference colours.

Fe-rich chlorite: R.I. ca. 1.65—1.69; —ve 2*V*, slow along, and normal interference colours.

According to Albee (1962), chlorite in amygdales and as veins in ultramafic rocks deviate from the above generalisations.

Orientation diagrams. *(001) section and cleavage fragments* (Fig. 72a): acute bisectrix figure; because of very low δ', sections may appear isotropic, and isogyres are difficult to observe; may display anomalous interference colours; pleochroism absent, and no cleavage visible.

(010) section (Fig.72b): flash figure; δ (very low to low) up to 0.015; often anomalous interference colours; good cleavage, fast along (+ve) or slow along (—ve); almost straight extinction (may be mottled); pleochroic with maximum absorption parallel to cleavage.

(100) section: almost identical to the (010) section.

Optic-axis figures obtained in basal or near-basal sections showing no cleavage or pleochroism — but isogyres usually difficult to observe.

Occurrence. Chlorite is widespread in metamorphic rocks of low grade, particularly chlorite-schists derived from pelites, and greenschists derived from basic igneous rocks. In both metamorphic and igneous rocks, chlorite is a common alteration product of pyroxene, amphibole and biotite, the composition of the chlorite being closely related to

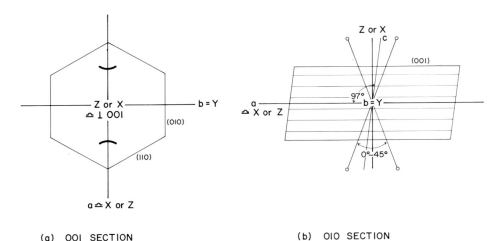

(a) OOI SECTION   (b) OIO SECTION

Fig. 72. Orientation diagrams for chlorite.

that of the original mineral. Chlorite is particularly characteristic of spilites and adinoles. Many amygdales in lavas are filled with chlorite, and chlorite is also a very common vein mineral, often associated with adularia and quartz. In sediments, chlorite is a very common detrital and authigenic mineral, often as mixed-layer structures, e.g. composite crystals made of chlorite and clay minerals. *Chamosite* is a name used for chlorite forming ooliths in sedimentary iron formations; some chamosite is, however, now known to be a clay mineral. Mn-rich and Cr-rich chlorites are found in Mn- and Cr-rich rocks respectively.

Distinguishing features (see Table II). The good single cleavage, green pleochroism, low δ and anomalous interference colours are distinctive. The length-fast character of +ve chlorites distinguishes these varieties from all other sheet silicates except prehnite. Chlorites with normal interference colours may be confused with serpentine, but serpentine generally has a lower R.I. and weaker pleochroism. Chloritoid may be mistaken for chlorite, but it has a higher R.I., and a larger $2V$ and extinction angle. Glauconite has a much higher δ than chlorite.

No. 37. SERPENTINE (−ve) $Mg_3[Si_2O_5](OH)_4$

Small amounts of Fe and Al are usually present.
As for the chlorites, an adequate characterisation of the serpentine polymorphs is only possible using X-rays. Three varieties are normally distinguished, *chrysotile*, *lizardite*, and *antigorite*. Chrysotile is usually fibrous, lizardite platy, and antigorite fibrous or platy. Lizardite has a high Fe^{3+}/Fe^{2+} ratio.

Monoclinic, β ca. 90°−93°. Fine-grained aggregates; fibrous (asbestiform) usually parallel to a, or platy with a perfect $\{001\}$ cleavage.

Colour in thin section: colourless to pale green.

Optical properties: biaxial −ve. $2V_x$ ca. = 20°−60°.
$\alpha = 1.532-1.570$, $\gamma = 1.545-1.584$.

δ = 0.004—0.017. X ca. $\perp$ (001) in antigorite and lizardite; X or Z parallel to fibre length in chrysotile.
The lowest R.I. and highest δ are associated with chrysotile, the highest R.I. and lowest δ with antigorite and lizardite.

Orientation diagrams. *Platy serpentine*: very similar to —ve chlorite. Sections displaying cleavage have straight extinction and are slow along. (001) sections show no cleavage, and provide acute bisectrix figures, but these are difficult to obtain due to the fine-grained nature of most serpentine. Interference colours are normal, and pleochroism weak or absent.
Fibrous serpentine: straight extinction, and slow or fast along. Figures impossible to obtain because of fine-grained size of fibres.

Occurrence. Serpentine forms by the hydrothermal alteration of ferromagnesium minerals below 400—500°C, particularly olivine and pyroxene. Serpentine replaces olivine as an irregular anastomosing network (Fig. 73), but more regular arrangements mimicking cleavage are formed from pyroxene and amphibole. Replacement is usually accompanied by the formation of small grains of opaque iron oxides. Ultramafic rocks, such as dunite, may be completely replaced, forming large rock masses of serpentinite.

Distinguishing features (see Table II). The occurrence, platy or fibrous nature, low δ and R.I. are characteristic. The generally lower R.I., weaker pleochroism and colour, and lack of anomalous interference colours distinguish serpentine from chlorite. Fibrous amphiboles have a higher R.I.

Fig. 73. Serpentine (view measures 1.5 mm × 1.0 mm). Crossed-polarised light.

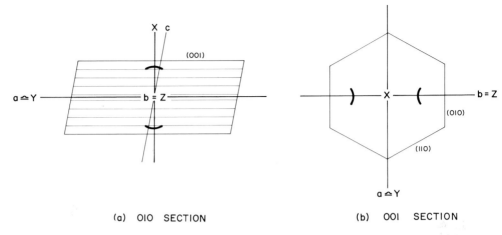

(a) OIO SECTION   (b) OOI SECTION

Fig. 74. Orientation diagrams for pyrophyllite.

No. 38. PYROPHYLLITE (—ve) $Al_4[Si_8O_{20}](OH)_4$

Monoclinic (may also be triclinic), $\beta = 100°$. Perfect $\{001\}$ cleavage. Subhedral tabular crystals parallel to $\{001\}$, or fine-grained aggregates, often as radiating groups.

Colour in thin section: colourless.

Optical properties: biaxial —ve. $2V_x = 53°—62°$.
$\alpha = 1.534—1.556$, $\beta = 1.586—1.589$, $\gamma = 1.596—1.601$.
$\delta = 0.045—0.062$. X ca. $= \perp (001)$, Y ca. $= a$, $Z = b$.

Orientation diagrams. *(010) section* (Fig. 74a): obtuse bisectrix figure; δ' (high) ca. 0.05; slow along good cleavage, and straight extinction.
(100) section: similar to (010) section, but flash figure.
(001) section (Fig. 74b): acute bisectrix figure; δ' (low) ca. 0.01; no cleavage visible; extinction may be wavy due to bending of crystals.
Optic-axis figures are obtained in sections oblique to (001) that display no cleavage.

Occurrence. Not common. Found in low-grade schists, and as a sericitic hydrothermal alteration product of feldspars, kyanite, and andalusite.

Distinguishing features (see Table II). Closely resembles muscovite and talc and other similar minerals. Muscovite and especially talc can be distinguished by their smaller $2V$, but with fine-grained material, chemical or X-ray tests may be necessary.

No. 39. TALC (—ve) $Mg_6[Si_8O_{20}](OH)_4$

Monoclinic, $\beta = $ ca. $100°$; more rarely triclinic. Perfect $\{001\}$ cleavage. Usually in massive aggregates of flaky crystals with the $\{001\}$ form dominating. Very soft, with a soapy feel in hand specimen.

Colour in thin section: colourless.

Optical properties: biaxial —ve. $2V_x = 0° - 30°$.
$\alpha = 1.539 - 1.550$, $\beta = 1.584 - 1.594$, $\gamma = 1.548 - 1.596$.
$\delta = 0.039 - 0.050$. X ca. $= \perp (001)$, Y ca. $= a$, $Z = b$.
Fe-rich varieties of talc are known with a higher R.I. (e.g. $\alpha = 1.580$, $\gamma = 1.615$).

Orientation diagrams. As for muscovite (No. 31), but has a smaller $2V$.

Occurrence. As a hydrothermal replacement product of ultramafic rocks, especially serpentinite, in veins and as massive bodies (soapstone), often associated with magnesite, tremolite, anthophyllite and chlorite. Also forms during the initial stages of the thermal metamorphism of impure dolomites.

Distinguishing features (see Table II). The occurrence, perfect cleavage, straight extinction (mottled), length-slow nature, and high δ are characteristic, but talc is easily confused with muscovite, pyrophyllite, and other similar minerals. Talc has a smaller $2V$ than pyrophyllite and most muscovite, and a lower R.I. than most muscovite with a small $2V$ (phengite). The softness and soapy feel of talc in hand specimen are distinctive.

No. 40. STILPNOMELANE (—ve)

$$(K,Na,Ca)_{0-1.4}(Fe,Mg,Al,Mn)_{5.9-8.2}[Si_8O_{20}](OH)_4(O,OH,H_2O)_{3.6-8.5}$$

Triclinic, $\beta = 96°$. Perfect $\{001\}$ and imperfect $\{010\}$ cleavages. As thin platy crystals dominated by $\{001\}$, and with pseudohexagonal outlines.

Colour in thin section. pale to dark yellow, green or brown, pleochroic with $Z = Y > X$; Fe^{3+}-rich varieties are usually very dark brown, and Fe^{2+}-rich varieties, dark green in colour.

Optical properties: biaxial —ve. $2V_x = 0° - 40°$ (usually ca. $0°$).
$\alpha = 1.543 - 1.634$, $\beta = \gamma = 1.576 - 1.745$.
$\delta = 0.033 - 0.111$. X ca. $= \perp (001)$, Y ca. $= b$, Z ca. $= a$.
R.I. and δ increase with Fe^{3+} content.

Orientation diagrams. *(001) section* (Fig. 75a): acute bisectrix figure, often pseudo-uniaxial; δ' (very low) ca. zero (appears isotropic); poor (010) cleavage parallel to O.A.P. may be visible; maximum absorption in this section, no pleochroism.
(010) section (Fig. 75b): flash figure; δ (high to very high) up to 0.111; good cleavage, straight extinction and length slow; pleochroism marked, with $Z > X$.
(100) section: similar to (010) section, but may show a poor cross-cleavage parallel to (010).

Occurrence. Relatively common as fine grains in low-grade schists and greenschists, especially those relatively rich in Fe and Mn. Also reported in xenoliths in igneous rocks.

Distinguishing features (see Table II). The strong pleochroism, straight extinction, habit, high δ and small $2V_x$ make stilpnomelane difficult to distinguish from green or brown biotite. However, the $\{001\}$ cleavage of stilpnomelane is less perfect than that of biotite, and there may be in addition a very poor $\{010\}$ cleavage. Mottled extinction is absent in stilpnomelane, and the flakes are not flexible as in biotite. Stilpnomelane may superficially resemble chlorite and chloritoid, but its δ is much higher.

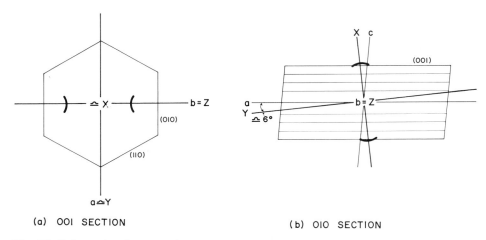

(a) OOI SECTION   (b) OIO SECTION

Fig. 75. Orientation diagrams for stilpnomelane.

No. 41. MARGARITE (−ve) $Ca_2Al_4[Si_4Al_4O_{20}](OH)_4$

A "brittle-mica" with a muscovite-like structure, but with Ca instead of K.

Monoclinic, β = 95°. Perfect $\{001\}$ cleavage. Subhedral platy crystals dominated by $\{001\}$. Twinning with $\{001\}$ composition plane.

Colour in thin section: colourless.

Optical properties: biaxial −ve. $2V_x$ = 40°−67°.
α = 1.630−1.638, β = 1.642−1.648, γ = 1.644−1.650.
δ = 0.012−0.014. $X \wedge \perp (001)$ = ca. 6°, $Y \wedge a$ = ca. 6°, $Z = b$.

Orientation diagrams. *(001) section* (Fig. 76a): acute bisectrix figure; δ' (very low) = 0.002; indistinct extinction positions; no cleavage visible.
(010) section (Fig. 76b): obtuse bisectrix figure; δ' (low) up to 0.012; good cleavage with slightly inclined extinction; slow along.
(100) section: similar to (010) section, but flash figure and straight extinction.
Optic-axis figures obtained in sections slightly oblique to (001).

Occurrence. Usually associated with corundum in emery deposits or veins, but also in some low-grade schists.

Distinguishing features (see Table II). Margarite may be confused with muscovite, talc and −ve chlorite. Muscovite and talc have a lower R.I., a higher δ, and more flexible cleavages; −ve chlorite is distinguished by its green colour, and usually by its lower δ.

No. 42. CLINTONITE (−ve) $Ca_2(Mg,Fe)_{4.6}Al_{1.4}[Si_{2.5}Al_{5.5}O_{20}](OH)_4$

A "brittle-mica" with a biotite-like structure, but with Ca instead of K.

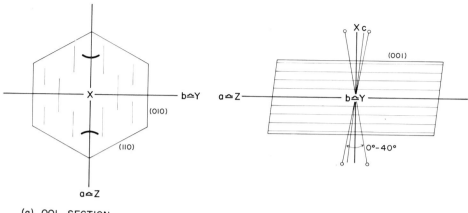

(a) 00I SECTION

(b) 0IO SECTION

Fig. 76. Orientation diagrams for margarite.

Monoclinic, β = 100°. Perfect {001} cleavage. Subhedral platy crystals with {001} dominating. Twinning with {001} composition plane.

Colour in thin section: colourless, pale green or brown, pleochroic with $Z = Y > X$.

Optical properties: biaxial —ve. $2V_x$ = 2°—40°.
α = 1.643—1.649, β = 1.655—1.662, γ = 1.655—1.663.
δ = 0.012—0.014. X ca. = $\perp$ (001), Y ca. = a, Z = b.
The related mineral *xanthophyllite* has X ca. = $\perp$ (001), Y = b, Z ca. = a.

Orientation diagrams. Similar to those for margarite (No. 41).

Occurrence. Not common. Found in chlorite-schists with talc, and, in metasomatically altered limestones.

Distinguishing features (see Table II). Distinguished from the micas by its lower δ and less flexible cleavage. —ve chlorite usually has a lower δ, and chloritoid a higher R.I.

No. 43. PREHNITE (+ve) $\qquad\qquad\qquad$ $Ca_2 Al[AlSi_3O_{10}](OH)_2$

Fe may substitute for Al.

Orthorhombic. Distinct {001} and poor {110} cleavages. Crystals are usually tabular parallel to (001) and are often arranged in radiating or sheaf-like aggregates; also prismatic parallel to c. Very fine lamellar twinning on {100} may be present.

Colour in thin section: colourless.

Optical properties: biaxial +ve. $2V_z$ = 65°—69°.
α = 1.611—1.632, β = 1.615—1.642, γ = 1.632—1.665.
δ = 0.021—0.035. X = a, Y = b, Z = c.

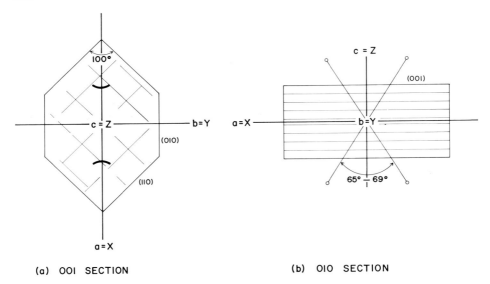

(a) OOI SECTION   (b) OIO SECTION

Fig. 77. Orientation diagrams for prehnite.

Orientation diagrams. *(001) section* (Fig. 77a): acute bisectrix figure; δ' (low) < 0.01; the poor $\{110\}$ cleavages may be visible.
(010) section (Fig. 77b): flash figure; δ (moderate to high) up to 0.035; distinct cleavage, straight extinction and fast along.
(100) section: straight extinction and fast along; similar to the (010) section, but a slightly lower δ', and provides an obtuse bisectrix figure.
Optic-axis figures are obtained on sections oblique to (001) showing no good cleavage.

Occurrence. Commonly found in veins and amygdales in basic volcanic rocks. Forms as a result of the hydrothermal alteration of Ca-bearing minerals such as plagioclase and amphibole. Also found in thermally metamorphosed impure carbonate rocks, and is a characteristic mineral in a wide variety of rocks affected by the low-grade prehnite– pumpellyite facies of regional metamorphism, and in veins cutting such rocks.

Distinguishing features (see Table II). The moderate R.I., high δ, good single cleavage, straight extinction, and +ve character are distinctive. It may be confused with muscovite, especially in metamorphic rocks, but is distinguished by its length-fast character, the lack of mottled extinction, and the less distinct cleavage.

No. 44. APOPHYLLITE (+ve or —ve) $KFCa_4[Si_8O_{20}]8H_2O$

Tetragonal. Perfect $\{001\}$ and imperfect $\{110\}$ cleavages. Often stubby euhedral prismatic crystals with $\{100\}$ well developed; also anhedral.

Colour in thin section: colourless.

Optical properties: uniaxial +ve, more rarely —ve.
$\omega = 1.5335\text{—}1.5445$, $\epsilon = 1.5352\text{—}1.5439$.
$\delta = 0.00\text{—}0.002$.

Interference colours are often anomalous.
Occasionally biaxial +ve, with complex twinning.

Occurrence. Fairly common in cavities and amygdales in basic volcanic rocks, usually associated with zeolites.

Distinguishing features. The R.I. (approximately the same as that of canada balsam), uniaxial character, anomalous interference colours, very low δ, and occurrence are characteristic. Associated zeolites have either a lower R.I. or a higher δ.

The clay minerals

Clay is used as a rock term, as a qualification of grain size, and also for a group of hydrous aluminium sheet-silicates characteristically found only in clays, argillites, soils, and related rocks. Sheet-silicates with a more widespread occurrence, such as chlorite and mica, may also be present in clay grain-size material.

Because of the very fine grain size of most clay minerals, they do not lend themselves to optical study. To study them properly requires the use of X-rays, differential thermal analysis, electron microscopy and chemistry. The description of the clay minerals given here will therefore be brief. Further details can be found in books such as Grim (1968).

No. 45. KAOLINITE GROUP

includes the following species:

No. 45A. Kaolinite $\qquad$ $Al_2[Si_2O_5](OH)_4$
No. 45B. Dickite, a polymorph of kaolinite.
No. 45C. Nacrite, a polymorph of kaolinite.
No. 45D. Halloysite, a hydrated form.

No. 46. SMECTITE GROUP

includes the following species:

No. 46A. Montmorillonite $\qquad$ $(\tfrac{1}{2}Ca,Na)_{0.66}(Al,Mg)_4[Si_8O_{20}](OH)_4 \cdot nH_2O$
No. 46B. Beidellite $\qquad$ $(\tfrac{1}{2}Ca,Na)_{0.66}Al_4[(Si,Al)_8O_{20}](OH)_4 \cdot nH_2O$
No. 46C. Nontronite $\qquad$ $(\tfrac{1}{2}Ca,Na)_{0.66}Fe_4^{3+}[(Si,Al)_8O_{20}](OH)_4 \cdot nH_2O$
No. 46D. Saponite $\qquad$ $(\tfrac{1}{2}Ca,Na)_{0.66}Mg_6[(Si,Al)_8O_{20}](OH)_4 \cdot nH_2O$
No. 46E. Hectorite $\qquad$ $(\tfrac{1}{2}Ca,Na)_{0.66}(Mg,Li)_6[Si_8O_{20}](OH)_4 \cdot nH_2O$

No. 47. ILLITE GROUP $\qquad$ $K_{1-1.5}Al_4[Si_{7-6.5}Al_{1-1.5}O_{20}](OH)_4 \cdot (+H_2O?)$

The status of this group is uncertain. Illite is essentially a mica (also known as hydro-muscovite). It differs from the micas in having less substitution of Al for Si and possibly being hydrated. Some "sericite" may be illite.

No. 48. VERMICULITE GROUP

$(Mg,Ca)_x(Mg,Fe)_6[(Si_{8-x},Al_x)O_{20}](OH)_4 \cdot 8H_2O$ where $x = 1-1.4$

TABLE III

Optical properties and crystal habits of the clay minerals

Mineral	α	γ	δ	2V and sign	Habit and other optics
Kaolinite (45A)	1.553–1.563	1.560–1.570	0.006–0.007	24°–50° (−ve)	{001} flakes; $X \wedge \perp (001) = 3^\circ$, Z ca. $= b$
Dickite (45B)	1.560–1.562	1.566–1.571	0.006–0.009	52°–80° (+ve)	{001} flakes; $X \wedge c = 15^\circ$–20°, $Z = b$
Nacrite (45C)	1.557–1.560	1.563–1.566	0.006	40° (−ve)	{001} flakes; $X \wedge (001) = 10^\circ$–$12^\circ$, $Z = b$
Halloysite (45D)	mean value $= 1.526$–1.556		0.002–0.001	?	elongate tubes
Montmorillonite and Beidellite (46A, B)	1.48 –1.57	1.50 –1.60	0.020–0.030	0°–30° (−ve)	usually flake aggregates; X ca. $= \perp (001)$
Nontronite (46C)	1.565–1.60	1.60 –1.640	0.035–0.040	moderate −ve	flakes or rods; yellow, brown, green, pleochroic
Saponite (46D)	1.480–1.490	1.510–1.525	0.030–0.035	moderate −ve	flakes; X ca. $= \perp (001)$
Hectorite (46E)	1.485	1.516	0.031	small −ve	laths; X ca. $= \perp (001)$
Illite (47)	1.545–1.63	1.57 –1.67	0.022–0.055	small −ve	flakes; X ca. $= \perp (001)$
Vermiculite (48)	1.525–1.56	1.545–1.585	0.020–0.030	small −ve	flakes; X ca. $= \perp (001)$; green brown, pleochroic
Allophane (49)	$n = 1.468$–1.512		isotropic		amorphous

No. 49. ALLOPHANE

A term used for amorphous clay material.

Crystallography and optical properties. Most of the clay minerals are monoclinic, though kaolinite is triclinic. Table III summarises the optical properties and crystal habits of the clay minerals — the data coming mainly from Grim (1968). It should be noted that properties such as R.I. may change during the drying of a sample, or as a result of absorption of immersion liquids during R.I. measurement. Identification is also hindered by the commonly mixed character of clay samples, and interlayering of different clay structures within individual crystals.

Occurrence. The factors governing the formation of particular clay-mineral species are exceedingly complex. There are, in general, three principal modes of occurrence: (1) as a result of hydrothermal alteration; (2) as a result of weathering, and in soils; (3) in sediments. Acid conditions appear to favour the formation of kaolinite, whereas alkaline conditions favour the smectites and illite. Illite is favoured by high concentrations of K, and may be derived from muscovite. Smectites are a characteristic alteration product of basic volcanic rocks. Vermiculite is often coarser grained than the other clays, and forms pseudomorphs after biotite. The clay minerals in a sediment reflect not only the parent soils and weathered rocks, but the conditions during and subsequent to deposition which may bring about diagenetic changes.

Distinguishing features. Apart from coarse-grained pseudomorphs of vermiculite after biotite, the clay minerals are characterised by their very fine grain size, and their occurrence in clays, soils, and weathered or altered rocks. Optical determination is not often possible. Members of the kaolinite group are distinguished from the other clay minerals by their low δ. Vermiculite and nontronite are coloured and pleochroic. Vermiculite differs from biotite in its weaker pleochroism and smaller δ. Flakes of the clay minerals are always length slow.

D. INOSILICATES

In the inosilicates, the $[SiO_4]$ tetrahedral groups are linked together to form chains. Minerals with this type of structure described here are: the pyroxenes, characterised by a single-chain structure with the formula $[SiO_3]$; wollastonite; pectolite; rhodonite; and the amphiboles, characterised by a double-chain structure with the formula $[Si_4O_{11}]$.

The pyroxenes

Crystallography

The pyroxenes may belong to either the monoclinic or the orthorhombic systems. In the monoclinic pyroxenes, the crystallographic angle β varies between $105°$ and $110°$. Most pyroxenes form rather stumpy prismatic crystals, though occasionally, as in the case of aegirine, the crystals are more elongate. Euhedral pyroxene crystals are characterised by an eight-sided cross-section, and all pyroxenes have two cleavages parallel to the c-axis that are almost at right angles to each other (Fig. 78).

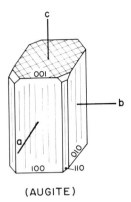

(AUGITE)

Fig. 78. Typical habit of augite.

The three following characteristic orientations may be observed in thin sections (Fig. 79): sections at right angles to the c-axis which display two cleavages (and an eight-sided cross-section in euhedral grains); prismatic sections displaying a single cleavage; prismatic sections parallel to (100) or

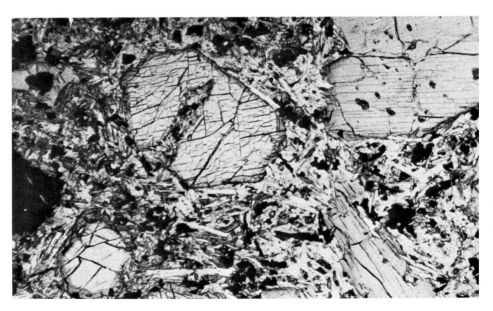

Fig. 79. Sections of augite: centre, eight-sided section with two cleavages; top right and bottom right, sections with traces of cleavage in one direction; bottom left, section with no trace of cleavage. Basalt, Pahau River, New Zealand (view measures 3.4 mm × 2.1 mm). Plane-polarised light.

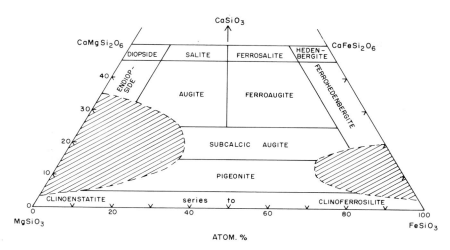

Fig. 80. Terminology of clinopyroxenes in the system $CaMgSi_2O_6-CaFeSi_2O_6-MgSiO_3-FeSiO_3$. The shaded areas represent areas of immiscibility and lacking naturally occurring examples. Simplified and redrawn with permission from Poldervaart and Hess (1951) and Deer et al. (1963).

(010) to which both cleavages are at too acute an angle to be always visible, especially in slightly thick sections.

Terminology

On the basis of the crystal system to which they belong, the pyroxenes are subdivided into clinopyroxenes (monoclinic) and orthopyroxenes.

Clinopyroxenes. The most important clinopyroxenes can be considered as members of the system $CaMgSi_2O_6-CaFeSi_2O_6-MgSiO_3-FeSiO_3$ (Fig. 80). Mg may be completely replaced by Fe^{2+} so that there is a continuous sequence of pyroxenes possible from $CaMgSi_2O_6$ to $CaFeSi_2O_6$, and from $MgSiO_3$ to $FeSiO_3$. There is more limited solid solution possible between the Ca-rich and Ca-poor pyroxenes, especially between the Mg-rich members. Subcalcic augite is not common, and usually forms under disequilibrium conditions of rapid crystallisation in volcanic rocks. Pyroxenes belonging to the series clinoenstatite—clinoferrosilite are seldom found in nature, and they are not described in this book; it is not certain whether they represent high-temperature or metastable forms of the common orthopyroxenes. Pigeonites are the common Ca-poor clinopyroxenes of intermediate Mg—Fe composition. They are metastably preserved in many volcanic rocks, but may invert on slow cooling to orthopyroxenes of similar composition.

The crystallisation of the Mg-rich pyroxenes in this system is represented diagrammatically in Fig. 81. Solid solution is incomplete between

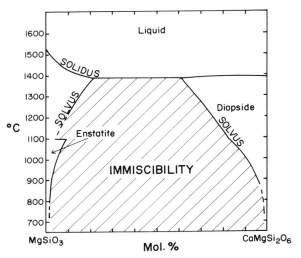

Fig. 81. Solidus and solvus curves in the system $CaMgSi_2O_6$—$MgSiO_3$. Redrawn with permission from Boyd and Schairer (1964).

$CaMgSi_2O_6$ and $MgSiO_3$, even at high temperatures, so that a melt of intermediate composition will crystallise both a Ca-rich and a Ca-poor pyroxene whose compositions are represented by the solvus curve.

The changes in composition of the two coexisting pyroxenes that crystallise under the influence of iron-enrichment during crystal fractionation in the well-known Skaergaard intrusion are illustrated in Fig. 82. The essential

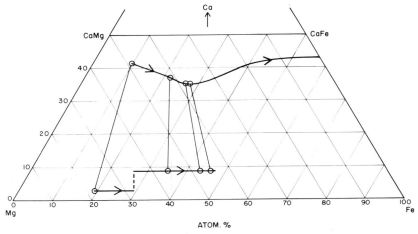

Fig. 82. Crystallisation trends and coexisting pyroxene pairs (tied circles) in the Skaergaard intrusion. Redrawn with permission from Brown and Vincent (1963).

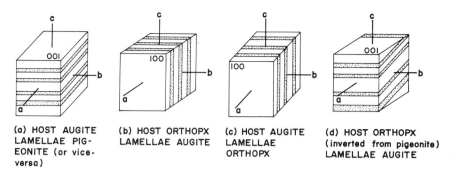

(a) HOST AUGITE LAMELLAE PIGEONITE (or vice-versa)   (b) HOST ORTHOPX LAMELLAE AUGITE   (c) HOST AUGITE LAMELLAE ORTHOPX   (d) HOST ORTHOPX (inverted from pigeonite) LAMELLAE AUGITE

Fig. 83. Exsolution-lamellae orientations in pyroxenes. Redrawn with permission from Poldervaart and Hess (1951). *Note:* b and c of lamellae and host are always parallel.

features are: initial crystallisation of a diopsidic augite and Mg-rich orthopyroxene; at a composition of about $En_{70}Fs_{30}$, pigeonite crystallises (with augite) in place of the orthopyroxene; with further fractionation, both the augite and pigeonite become more Fe-rich until eventually pigeonite ceases to crystallise. Very similar compositional trends are known from a large number of rock suites similarly affected by iron enrichment. In some volcanics, a different trend from diopside towards subcalcic augite and pigeonite has been observed (Kuno, 1955), this being regarded as typical of rapid cooling from high temperatures. Nakamura and Kushiro (1970) have described rocks in which the three pyroxenes pigeonite, orthopyroxene and augite all coexist.

Other less common clinopyroxenes that cannot be represented by the chemical substitutions shown in Fig. 80 include omphacite, fassaite, aegirine, jadeite, and spodumene. Omphacite resembles diopside, but is distinguished by substitution of Na for some of the Ca, and Al for some of the Mg. Fassaite also resembles diopside, but there is a significant replacement of both Mg and Si by Al. Aegirine has the formula $NaFe^{3+}[Si_2O_6]$ and aegirine-augite has a composition intermediate between this and members of the diopside—hedenbergite series; crystallisation trends in this system are discussed by Gomes et al. (1970). Jadeite has the formula $NaAl[Si_2O_6]$, and spodumene is a Li-pyroxene with the composition $LiAl[Si_2O_6]$.

Orthopyroxenes. The series of pyroxenes from clinoenstatite to clinoferrosilite represented in Fig. 80 occur only rarely, and their place is taken in nature by a series of orthopyroxenes from enstatite to orthoferrosilite. Pigeonites commonly invert to orthopyroxenes on cooling. The nomenclature of the orthopyroxenes is given in Fig. 84.

Exsolution phenomena

The two coexisting pyroxenes that initially crystallise in basic igneous rocks (Fig. 82) may be metastably preserved on rapid cooling. However, the composition of each pyroxene will change during slow cooling in a way prescribed by the solvus curve of Fig. 81. A Ca-poor pyroxene will become even more Ca-poor, and in so doing will exsolve a Ca-rich pyroxene as exsolution lamellae (see also Fig. 102 and the accompanying text for details of a similar process in the feldspars). The Ca-rich pyroxene that originally crystallises will exsolve a Ca-poor pyroxene. Poldervaart and Hess (1951) have shown that a clinopyroxene exsolved from a clinopyroxene (e.g. pigeonite from augite and vice-versa) will form exsolution lamellae parallel to (001); a clinopyroxene exsolved from orthopyroxene (or vice-versa) will form exsolution lamellae parallel to (100). The various possible orientations of the lamellae are shown diagrammatically in Fig. 83. Pigeonite initially exsolves augite parallel to (001), but may then invert to orthopyroxene and continue to exsolve augite parallel to (100); the relict (001) lamellae are preserved on an irrational plane in the orthopyroxene. Similarly, augite initially exsolves pigeonite, but on further cooling exsolves orthopyroxene.

No. 50. ORTHOPYROXENE (+ve and —ve)   (Mg,Fe)[SiO$_3$]

The terminology of the orthopyroxenes is given in Fig. 84.

Orthorhombic. Distinct {210} cleavages at 88° to each other. Parting and exsolution lamellae commonly developed parallel to {100} (Fig. 83). Often euhedral stubby pris-

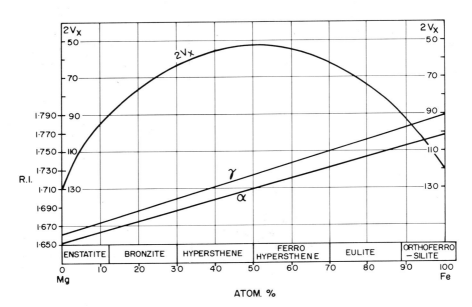

Fig. 84. Terminology of the orthopyroxenes, and variation of R.I. and 2V with composition. Redrawn with permission from Leake (1968).

matic crystals with eight-sided cross-sections; also anhedral. Simple or lamellar twins with {100} composition planes common.

Colour in thin section: colourless, or pale-pink and green pleochroic with X = pink, Y = pale yellow, Z = green.

Colour in detrital grains: as above but deeper colours.

Optical properties: biaxial +ve or —ve. $2V_x$ = 45°—128°.
α = 1.651—1.769, β = 1.653—1.771, γ = 1.658—1.788.
δ = 0.007—0.019. X = b, Y = a, Z = c.
R.I. and δ increase and $2V_x$ varies with Fe content as shown in Fig. 84.

Orientation diagrams. *(001) section* (Fig. 85a): acute or obtuse bisectrix figure; δ' (very low or low) up to 0.010; symmetrical extinction to two distinct cleavages almost at right angles to each other; eight-sided sections typical; may display twinning and exsolution lamellae parallel to (100).
(100) section (Fig. 85b): flash figure; δ usually low but may be as high as 0.019 in orthoferrosilite; straight extinction, but cleavage not always seen; may show marked pleochroism; if inverted from pigeonite may display exsolution lamellae on plane parallel to b but oblique to (001).
(010) sections are similar to (100) sections but provide bisectrix figures, and may display augite exsolution lamellae parallel to (100) or oblique to (001) if inverted from pigeonite. *Sections displaying a sharp single cleavage* are oblique to the optical directions.
Optic-axis figures are obtained in sections oblique to the cleavages.

Occurrence. Mg-rich orthopyroxenes are characteristic of ultramafic rocks such as pyroxenites and harzburgite. Orthopyroxenes of varying Mg/Fe content are common in a wide range of basic igneous rocks, especially norite and andesite. Orthopyroxenes are also found in charnockites, granulites, and pelitic rocks thermally metamorphosed at high

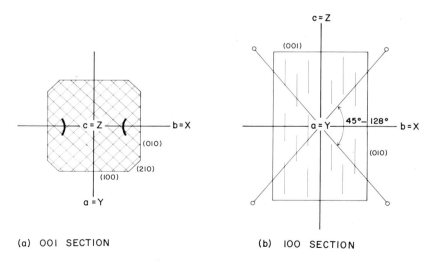

(a) 001 SECTION   (b) 100 SECTION

Fig. 85. Orientation diagrams for orthopyroxene.

grades. Orthoferrosilite occurs in some metamorphosed Fe-rich sediments. The ortho-pyroxenes are commonly altered to serpentine, chlorite, or uralite (amphibole).

Distinguishing features. The two cleavages at right angles, straight extinction in prismatic sections, high relief, and low δ are characteristic. It may be difficult to distinguish from augite, especially when present in small quantities with augite. The best way to identify the orthopyroxene in these circumstances is to obtain numerous optic-axis figures. Augite always has a moderate +ve $2V$ whereas the orthopyroxene most commonly has a large $2V$ or is —ve. It may occasionally be confused with andalusite, especially in detrital grains, but andalusite is length fast, and euhedral crystals are different in shape.

No. 51A. DIOPSIDE—HEDENBERGITE (+ve) $Ca(Mg,Fe)[Si_2O_6]$

No. 51B. AUGITE (+ve) $(Ca,Mg,Fe,Ti,Al)_2[(Si,Al)_2O_6]$

No. 51C. SUBCALCIC AUGITE (+ve) $(Mg,Fe,Ca,Al)_2[(Si,Al)_2O_6]$

No. 51D. OMPHACITE (+ve) $(Ca,Na,Mg,Fe,Al)_2[Si_2O_6]$

No. 51E. FASSAITE (+ve) $Ca(Mg,Fe,Al)[(Si,Al)_2O_6]$

The composition of these closely related minerals is discussed in the introductory section on the pyroxenes.

Monoclinic, $\beta = 105°$—$106°$. Distinct $\{110\}$ cleavages at $87°$ to each other. Parting and exsolution lamellae of orthopyroxene commonly developed parallel to $\{100\}$ (Fig. 83). Often euhedral stubby prismatic crystals with eight-sided cross-sections; also anhedral. Simple or lamellar twinning with $\{100\}$ composition planes common. Oscillatory, hour-glass, and sector-zoning are often present, especially in Ti-rich titanaugite.

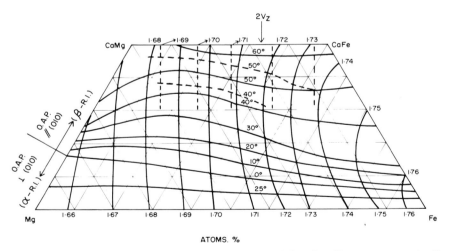

Fig. 86. Variation of $2V$ and R.I. with composition in clinopyroxenes in the system $CaMgSi_2O_6$—$CaFeSi_2O_6$—$MgSiO_3$—$FeSiO_3$. Redrawn with permission from Hess (1949, plate I), Muir (1951), and (dashed lines) Brown and Vincent (1963).

Colour in thin section: diopside, colourless; hedenbergite, pale green or green-brown; augite may be colourless or pale green, but a flesh colour is particularly common; titanaugite has a characteristic purple-brown colour; omphacite and fassaite are usually pale green; usually weak pleochroism.

Colour in detrital grains: as above but deeper colours.

Optical properties: biaxial +ve. $2V_z = 25°—83°$.
$\alpha = 1.659—1.743$, $\beta = 1.670—1.750$, $\gamma = 1.688—1.772$.
$\delta = 0.018—0.033$. $X \wedge a = 20°—34°$, $Y = b$, $Z \wedge c = 35°—48°$ (higher values up to 70° have been reported for omphacite).
R.I. increases with Fe content. $2V$ increases with Ca and Na content. Fig. 86 illustrates the variation of R.I. and $2V$ for the diopside—hedenbergite series, augite and subcalcic augite (refer to Fig. 80 for nomenclature). Fassaite has $2V_z = 51°—65°$, and for omphacite, $2V_z = 58°—83°$. The extinction angle $Z \wedge c$ increases with Fe content (Hess, 1949), but correlation of this angle with chemistry is not good.

Orientation diagrams. *Section $\perp c$* (Fig. 87a): close to an optic-axis figure; δ' low or very low; extinction symmetrical to two good cleavages almost at 90° to each other, but extinction may be indistinct due to proximity of an optic axis; eight-sided section characteristic; may display twinning and exsolution lamellae of orthopyroxene // (100).
(010) section (Fig. 87b): flash figure; δ (moderate) up to 0.033; inclined extinction; the trace of the prismatic cleavage is not always visible; twinning and exsolution lamellae of orthopyroxene // (100), and lamellae of pigeonite // (001) may be visible.
(100) section (Fig. 87c): close to an optic-axis figure; δ' low or very low; straight extinction; trace of cleavage is not always visible; exsolution lamellae of pigeonite // (001) may be visible.
Cleavage fragments (and some sections showing a good single cleavage) lie parallel to (110). The extinction angle $Z' \wedge c$ is smaller by 10° or so than the angle $Z \wedge c$ in (010) sections.

Occurrence. Diopside—hedenbergite occurs in metamorphic rocks, diopside being characteristic of thermally metamorphosed and metasomatised carbonate rocks. Hedenbergite is formed in metamorphosed Fe-rich sediments. Members of the diopside—hedenbergite series also occur in basic igneous rocks, hedenbergite in syenites and more rarely granites. Augite is the typical pyroxene of many igneous rocks, and is an essential mineral of basalts and gabbros. Subcalcic augite is occasionally found in basalts and andesites, sometimes as the outer zone of zoned augite crystals (Kuno, 1955). Omphacite is the typical pyroxene of eclogites and associated rocks. Fassaite is found in metamorphosed limestones in association with spinel. The common alteration product of diopside and common augite is uralite (tremolite, actinolite or hornblende).

Distinguishing features. The clinopyroxenes described above are distinguished by their two cleavages at 87°, their moderate δ, the moderate extinction angle in (010) sections, and their +ve character. They are not easily confused with any other mineral. However, it is easy to overlook small quantities of orthopyroxene (usually —ve or a large $2V$) associated with augite. Individual species of clinopyroxene are less easy to distinguish without accurate $2V$, R.I. measurements and chemical analysis. Omphacite and fassaite are characterised by a higher $2V$ and a pale-green colour. Diopside is colourless, whereas augite is commonly flesh coloured. Titaniferous augite is purple-brown in colour. Subcalcic augite has a small but larger $2V$ than pigeonite (No. 52). The related pyroxenes jadeite (No. 54) and spodumene (No. 55) have distinctive occurrences, and aegirine (No. 53) is green pleochroic, and fast along.

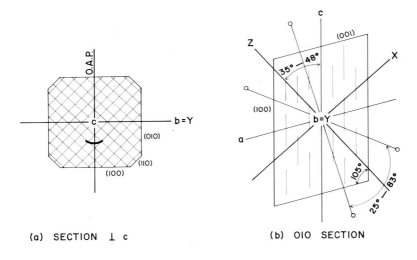

(a) SECTION ⊥ c   (b) OIO SECTION

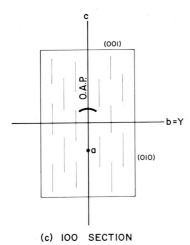

(c) IOO SECTION

Fig. 87. Orientation diagrams for the clinopyroxenes 51A—E.

No. 52. PIGEONITE (+ve) $(Mg,Fe,Ca)(Mg,Fe)[Si_2O_6]$

See Fig. 80 for the compositional boundaries of pigeonite.

Monoclinic, $\beta = 108°$. Distinct $\{110\}$ cleavages at 87° to each other. Usually as very small crystals in the groundmass of volcanic rocks. Twinning with $\{100\}$ composition plane.

Colour in thin section: colourless or pale pink or green; pleochroism weak or absent.

Optical properties: biaxial +ve. $2V_z = 0°-30°$.
$\alpha = 1.682-1.722$, $\beta = 1.684-1.722$, $\gamma = 1.705-1.751$.
$\delta = 0.023-0.029$. $X = b$ or $X \wedge a = 19°-26°$, $Y \wedge a = 19°-26°$ or $Y = b$, $Z \wedge c = 37°-44°$.
R.I. increases with Fe content. For the variation of R.I. and $2V$ with composition see Fig. 86.

Orientation diagrams. Similar to augite (Fig. 87), but O.A.P. is most commonly $\perp (010)$, and because of the very small $2V$ about Z, optic-axis figures are not obtained on sections $\perp c$ or $// (100)$.

Occurrence. Restricted to rapidly chilled lavas, usually in the groundmass of andesites and dacites. Also as exsolution lamellae in some augite. Pigeonite that has inverted to ortho-pyroxene may be detected by the presence of exsolution lamellae of augite on the relict (001) plane of pigeonite.

Distinguishing features. Very similar to the other clinopyroxenes (No. 51), but distinguished from them by its very small $2V$.

No. 53A. AEGIRINE (ACMITE) (−ve) $NaFe^{3+}[Si_2O_6]$

No. 53B. AEGIRINE-AUGITE (+ve and −ve) $(Na,Ca)(Fe^{3+},Fe^{2+},Mg,Al)[Si_2O_6]$

Monoclinic, $\beta = $ ca. $107°$. Distinct $\{110\}$ cleavages at $87°$ to each other. Aegirine is usually euhedral or subhedral with a very elongate or acicular prismatic habit. Aegirine-augite forms less elongate crystals. Twinning with $\{100\}$ composition planes. Zoned crystals are common.

Colour in thin section: pale to dark yellowish green; pleochroism weak to strong with $X > Y > Z$; acmite is distinguished from aegirine on the basis of colour, and is brown or brownish-green in thin section.

Optical properties: biaxial −ve or +ve. $2V_x = 60°-70°$ (aegirine), $2V_x = 70°-110°$ (aegirine-augite).
$\alpha = 1.700-1.776$, $\beta = 1.710-1.820$, $\gamma = 1.730-1.836$.
$\delta = 0.030-0.060$. $X \wedge c = 0°-20°$, $Y = b$, $Z \wedge a = 7°-37°$.
The variation in optical properties with composition is shown in Fig. 88.

Orientation diagrams. *Section $\perp c$* (Fig. 89a): acute bisectrix figure for aegirine obtuse bisectrix figure for some aegirine-augite; δ' (moderate) up to 0.020; symmetrical extinction to two distinct cleavages almost at right angles to each other; eight-sided sections typical.
(010) section (Fig. 89b): flash figure; δ (high) = 0.030—0.060, highest for aegirine; small extinction angle $X \wedge c$ and fast along; cleavage oblique to section and may not be visible; may show marked pleochroism.
(100) section: similar to the (010) section, but will provide an obtuse or acute bisectrix figure; straight extinction and fast along.

Occurrence. The typical pyroxene of alkaline igneous rocks including syenites, nepheline-syenites, trachytes and phonolites. Occasionally occurs in metamorphosed Na-rich rocks (e.g. spilite) in association with glaucophane or riebeckite.

Distinguishing features. The pyroxene habit and cleavages are distinctive but may be

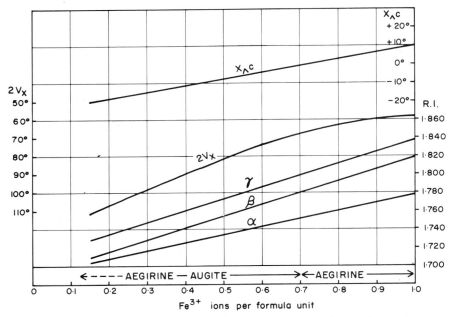

Fig. 88. Variation of $2V$, $X \wedge c$, and R.I. with composition in aegirine-augite and aegirine. Redrawn with permission from Deer et al. (1966).

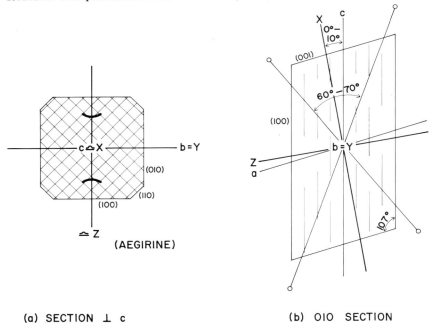

(a) SECTION ⊥ c

(b) OIO SECTION

Fig. 89. Orientation diagrams for aegirine.

difficult to observe in strongly prismatic crystals. Distinguished from other pyroxenes by its colour, prismatic habit, and fast-along character. The fast-along character distinguishes aegirine and aegirine-augite from similarly coloured amphiboles.

No. 54. JADEITE (+ve) $NaAl[Si_2O_6]$

Monoclinic, $\beta = 107°$. Distinct $\{110\}$ cleavages at 87° to each other. Usually anhedral and granular or fibrous. Twinning with $\{100\}$ composition planes.

Colour in thin section: colourless.

Optical properties: biaxial +ve. $2V_z = 67° - 86°$.
$\alpha = 1.640-1.681$, $\beta = 1.645-1.684$, $\gamma = 1.652-1.692$.
$\delta = 0.008-0.015$. $X \wedge a = 13°-38°$, $Y = b$, $Z \wedge c = 30°-55°$.

Orientation diagrams. Similar to Fig. 87, but a lower δ; euhedral crystals not common.

Occurrence. Restricted to metamorphic rocks formed under unusually high pressures. Common associates are albite, lawsonite and glaucophane. Precious jade is composed of jadeite, and forms in association with the metamorphism of ultramafic rocks.

Distinguishing features. The typical pyroxene cleavages and the restricted occurrence are distinctive. It may be confused with omphacite in eclogites but differs from omphacite (and other clinopyroxenes) in its lower δ. It may be confused with some amphiboles, but differs in its larger extinction angle, its pyroxene cleavages, and the complete lack of colour.

No. 55. SPODUMENE (+ve) $LiAl[Si_2O_6]$

Monoclinic, $\beta = 110°$. Distinct $\{110\}$ cleavages at 87° to each other. Often euhedral as very large tabular crystals with $\{100\}$ prominent and elongate parallel to c. Parting parallel to $\{100\}$. Twins with $\{100\}$ composition plane common.

Colour in thin section: colourless.

Optical properties: biaxial +ve. $2V_z = 55°-70°$.
$\alpha = 1.648-1.663$, $\beta = 1.655-1.669$, $\gamma = 1.662-1.679$.
$\delta = 0.014-0.027$, $X \wedge a = 2°-6°$, $Y = b$, $Z \wedge c = 22°-26°$.

Orientation diagrams. Similar to Fig. 87, but extinction angle $Z \wedge c$ is smaller.

Occurrence. Characteristic of Li-rich granite-pegmatites in which the spodumene crystals may be very large. Alters to micas and albite or clay minerals.

Distinguishing features. The pyroxene habit and cleavages, and occurrence are distinctive. Differs from other clinopyroxenes in its smaller extinction angle $Z \wedge c$.

No. 56. WOLLASTONITE (−ve) $Ca[SiO_3]$

Parawollastonite is a polymorph of wollastonite.

Triclinic, $\alpha = $ ca. 90°, $\beta = 95°$, $\gamma = 103°$ (parawollastonite is monoclinic with $\beta = 95°$).

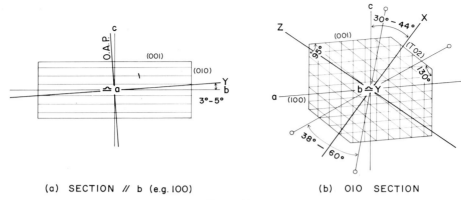

(a) SECTION // b (e.g. 100)   (b) 010 SECTION

Fig. 90. Orientation diagrams for wollastonite.

Perfect {100} and distinct {001} and {$\bar{1}02$} cleavages. Crystals are often euhedral to subhedral as tablets parallel to {100} and/or elongated parallel to b. Multiple twinning with {100} composition planes and [010] twin axis common.

Colour in thin section: colourless.

Optical properties: biaxial —ve. $2V_x = 38°—60°$.
$\alpha = 1.616—1.640$, $\beta = 1.627—1.650$, $\gamma = 1.631—1.653$.
$\delta = 0.013—0.015$. $X \wedge c$ = ca. $30°—44°$, $Y \wedge b = 3°—5°$ ($Y = b$ for parawollastonite), $Z \wedge a$ = ca. $35°—45°$.

Orientation diagrams. *Sections parallel to b and the length of crystals* (Fig. 90a): may be bisectrix, optic-axis, or off-centred figures, but all indicate the O.A.P. is across the good cleavage traces and length of crystals; δ' (low or very low) up to 0.012; straight or inclined extinction up to 5° (always straight for parawollastonite) and may be either length fast or slow.
(010) section (Fig. 90b): flash figure; δ (low) up to 0.015; three cleavages may be visible; inclined extinction to all cleavages; this section may not be easily observed in fibrous wollastonite.

Occurrence. Wollastonite (and much less commonly parawollastonite) occurs in thermally metamorphosed and metasomatised carbonate rocks. It also occurs more rarely in alkaline undersaturated igneous rocks such as ijolites and phonolites.

Distinguishing features. The elongate crystals, the O.A.P. across the cleavages, the —ve character and occurrence are distinctive. May be confused with tremolite, pectolite, and diopside, all of which commonly occur with wollastonite. However, the low δ and the position of the O.A.P. distinguishes wollastonite from all three. In addition, pectolite and diopside are +ve, and tremolite has a larger $2V$.

No. 57. PECTOLITE (+ve) $Ca_2NaH[SiO_3]_3$

Triclinic, $\alpha = 90°$, $\beta = 95°$, $\gamma = 102°$. Perfect {100} and {001} cleavages. Crystals elongate or fibrous parallel to b.

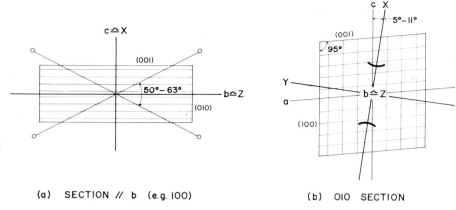

(a) SECTION // b (e.g. 100) (b) 010 SECTION

Fig. 91. Orientation diagrams for pectolite.

Colour in thin section: colourless.

Optical properties: biaxial +ve. $2V_z = 50° - 63°$.
$\alpha = 1.595 - 1.610$, $\beta = 1.605 - 1.615$, $\gamma = 1.632 - 1.645$.
$\delta = 0.030 - 0.038$. $X \wedge c = 5° - 11°$, $Y \wedge a = 10° - 16°$, Z is close to b.

Orientation diagrams. *Sections parallel to b and length of crystals* (Fig. 91a): flash or obtuse bisectrix figure; δ' (moderate to high) up to 0.038; almost straight extinction; trace of cleavage parallel to b often visible.
(010) section (Fig. 91b): acute bisectrix figure; δ' (very low) up to 0.005; inclined extinction to two cleavages at 95° to each other.

Occurrence. Most commonly found in hydrothermal veins and in cavities in basic igneous rocks. Also found in thermally metamorphosed carbonate rocks, and more rarely in undersaturated alkaline igneous rocks such as nepheline-syenite.

Distinguishing features. The elongate crystals, length-slow character, high δ and occurrence are characteristic. May be associated with prehnite and wollastonite, both of which superficially resemble pectolite. However, prehnite is length fast, and wollastonite is sometimes length fast, is —ve, and has a lower δ.

No. 58. RHODONITE (+ve) (Mn,Fe,Ca)[SiO₃]

Triclinic, $\alpha = 86°$, $\beta = 93°$, $\gamma = 111°$. Perfect {100} and {001} and distinct {010} cleavages. Euhedral to anhedral crystals, commonly tabular parallel to {010} . (N.B. crystal planes are indexed in a new way to be compatible with unit cell information.)

Colour in thin section: colourless to pale pink; pleochroism weak.

Optical properties: biaxial +ve. $2V_z = 61° - 76°$.
$\alpha = 1.711 - 1.738$, $\beta = 1.716 - 1.741$, $\gamma = 1.724 - 1.751$.
$\delta = 0.011 - 0.014$. $X \wedge c = $ ca. 45°, $Z \wedge b = $ ca. 20°.

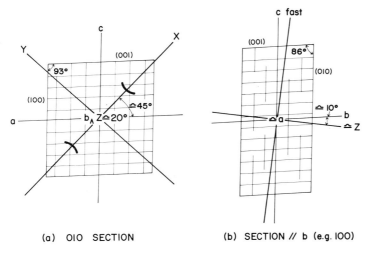

(a) OlO SECTION   (b) SECTION // b (e.g. lOO)

Fig. 92. Orientation diagrams for rhodonite.

Orientation diagrams. *(010) section* (Fig. 92a): since $b \wedge Z = 20°$, sections displaying two sharp cleavages at 93° provide off-centred acute bisectrix figures; δ' (very low) up to 0.005; almost symmetrical extinction.

Sections parallel to b (Fig. 92b): figures vary from flash to off-centred obtuse bisectrix figures; δ' (low) up to 0.014; may display perfect {001} or {100} cleavages and less distinct cross-cleavage {010} ; generally inclined extinction, and the slow direction is across the length of tabular crystals.

Occurrence. Found in ore deposits of manganese (often associated with metasomatic activity). Alters to pyrolusite (black MnO_2) or rhodochrosite.

Distinguishing features. The association with Mn-ore deposits, the pale-pink colour and triclinic nature are characteristic. It may be confused with clinopyroxene, but its lower δ is distinctive. The R.I. and δ are similar to those of kyanite, but the cleavages of kyanite are much more distinct.

The amphiboles

The amphiboles are a group of minerals characterised by the linkage of the $[(Si,Al)O_4]$ tetrahedra to form double chains of general composition $[(Si,Al)_4 O_{11}]_n$. This structure is reflected in two cleavages parallel to the *c*-axis at an angle of approximately 54° to each other. The structure allows very extensive ionic replacement, and the resulting variation in chemistry is so great that no completely satisfactory classification has been devised. The particular species and series described here are those most generally recognised, and they are listed below according to the threefold grouping of Ernst (1968).

Iron—magnesium amphiboles

No. 59A. ANTHOPHYLLITE　　　　　　$(Mg,Fe^{2+})_7[Si_8O_{22}](OH,F)_2$

No. 59B. GEDRITE　　　　　　$(Mg,Fe^{2+})_5Al_2[Si_6Al_2O_{22}](OH,F)_2$

No. 60A. CUMMINGTONITE　　　　　　$(Mg,Fe^{2+})_7[Si_8O_{22}](OH)_2$

which forms a series with:

No. 60B. GRUNERITE　　　　　　$Fe_7^{2+}[Si_8O_{22}](OH)_2$

Calcic amphiboles

No. 61A. TREMOLITE　　　　　　$Ca_2Mg_5[Si_8O_{22}](OH,F)_2$

which forms a series with:

No. 61B. FERRO-ACTINOLITE　　　　　　$Ca_2Fe_5^{2+}[Si_8O_{22}](OH,F)_2$

Intermediate members are called *actinolite*.

No. 62A. HORNBLENDE

composed of varying proportions of:

No. 62B. EDENITE　　　　　　$Ca_2Na(Mg,Fe^{2+})_5[Si_7AlO_{22}](OH,F)_2$

No. 62C. TSCHERMAKITE　　　　　　$Ca_2(Mg,Fe^{2+})_3Al_2[Si_6Al_2O_{22}](OH,F)_2$

No. 62D. PARGASITE　　　　　　$Ca_2NaMg_4Al[Si_6Al_2O_{22}](OH,F)_2$

No. 62E. FERRO-HASTINGSITE　　　　　　$Ca_2NaFe_4Al[Si_6Al_2O_{22}](OH,F)_2$

Hastingsite is an intermediate member of the pargasite—ferro-hastingsite series.

No. 63. OXYHORNBLENDE　　　$Ca_2(Na,K)(Mg,Fe^{2+},Fe^{3+},Al)_5[Si_6Al_2O_{22}](O,OH)_2$

Also known as *lamprobolite* or *basaltic hornblende*.
Kaersutite is a Ti-rich oxyhornblende.

No. 64. BARKEVIKITE　　　$Ca_2(Na,K)(Fe^{2+},Mg,Fe^{3+},Mn)_5[Si_6Al_2O_{22}](OH)_2$

Sodic amphiboles

No. 65A. GLAUCOPHANE　　　　　　$Na_2Mg_3Al_2[Si_8O_{22}](OH)_2$

which forms a series with:

No. 65B. RIEBECKITE　　　　　　$Na_2Fe_3^{2+}Fe_2^{3+}[Si_8O_{22}](OH)_2$

No. 66A. ECKERMANNITE $Na_3Mg_4Al[Si_8O_{22}](OH,F)_2$

   which forms a series with:

No. 66B. ARFVEDSONITE $Na_3Fe_4^{2+}Al[Si_8O_{22}](OH,F)_2$

No. 67. KATAPHORITE $Na_2Ca(Mg,Fe^{2+})_4Fe^{3+}[Si_7AlO_{22}](OH,F)_2$

As a group, the amphiboles are distinguished from other minerals by their two cleavages at $54°$, their moderate to high R.I., moderate to high δ, and their biaxial character. Many amphiboles are strongly coloured and pleochroic. Identification of individual species is less easy since there are considerable overlaps in their properties. In order to compare them more easily and avoid unnecessary repetition, the details are compiled below under the three main headings — crystallography, optical properties, and occurrence. The distinguishing features of each species are then itemised to assist identification.

Crystallography

All the amphiboles are monoclinic except anthopyllite and gedrite which are orthorhombic. For the monoclinic members, the crystallographic angle β varies within the small range $103°-109°$.

Crystal habit. Crystal habit varies from stubby prismatic (not common) to elongate prismatic (very common) to fibrous or asbestiform (common). Amphiboles which may be fibrous or asbestiform include anthophyllite, gedrite, and members of the cummingtonite—grunerite, tremolite—ferroactinolite, and glaucophane—riebeckite series. Asbestiform riebeckite, cummingtonite, and grunerite are known respectively as *crocidolite, montasite,* and *amosite.*

Euhedral prismatic crystals are generally four- or six-sided in cross-section with the {110} form dominating, with or without {010} (Fig. 93).

Cleavage. All amphiboles have perfect {110} (or {210} for orthorhombic amphibole) cleavages. The angle between the two cleavages is always in the range $54°-56°$, and b bisects the acute angle. Therefore, three principal types of section are encountered in thin section (Figs. 94, 97 and 98): sections approximately at right angles to the c-axis which display two cleavages; sections parallel to c and close to (010) or (110) which display the traces of either or both cleavages in one direction parallel to the c-axis; sections parallel to (100) which display *no* cleavage since both cleavages are at too acute an angle to the section to be visible.

Twinning. Simple or lamellar twinning with {100} composition planes is

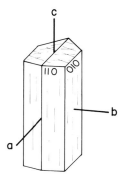

Fig. 93. Typical habit of a prismatic amphibole crystal.

common in all the monoclinic amphiboles, but is not possible in orthorhombic anthophyllite and gedrite.

Optical properties

The range of properties for particular species cannot be simply correlated

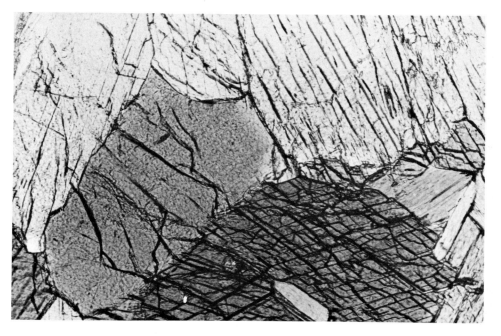

Fig. 94. Sections of hornblende: bottom right, section with two cleavages; centre left,
section with no cleavage visible; top left and right, sections with traces of cleavage in one
direction. Rotoroa Igneous Complex, New Zealand (view measures 0.7 mm × 0.45 mm).
Plane-polarised light.

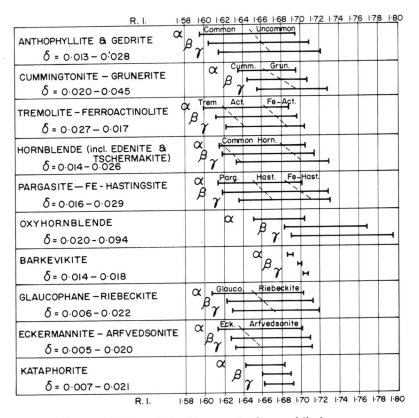

Fig. 95. Ranges of R.I. and birefringence in the amphiboles.

with chemistry, especially in the hornblende series. Ranges given by different authorities vary; those given here are taken mainly from Deer et al. (1966), with additional data from other sources.

Refractive index and birefringence (Fig. 95). Refractive indices within the various series increase with iron content. However, a simple correlation of R.I. with chemistry is not possible.

2V and the position of the O.A.P. (Fig. 96). Most amphiboles are −ve, but Fe-rich anthophyllite, most gedrite, cummingtonite, and some hornblende are +ve. As with R.I., a simple correlation of 2V with chemistry is not possible. 2V is difficult to determine in some of the sodic amphiboles because of their high dispersion and intense colour. The O.A.P. is parallel to (010) except in riebeckite, arfvedsonite, and Fe-kataphorite, in which the O.A.P. is perpendicular to (010).

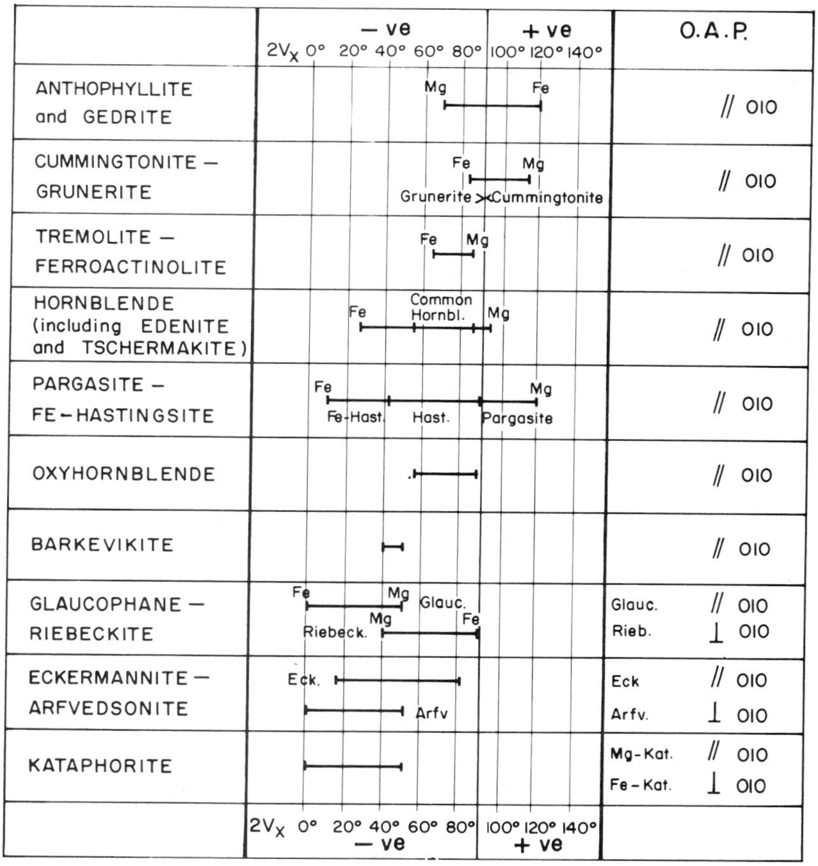

	− ve 2V$_X$ 0° 20° 40° 60° 80°	+ ve 100° 120° 140°	O.A.P.
ANTHOPHYLLITE and GEDRITE	Mg ⊢━━┿	Fe ━━┥	// 010
CUMMINGTONITE – GRUNERITE	Fe ⊢━┿	Mg ━━┥ Grunerite ⊁ Cummingtonite	// 010
TREMOLITE – FERROACTINOLITE	Fe ⊢━┥ Mg		// 010
HORNBLENDE (including EDENITE and TSCHERMAKITE)	Fe ⊢━┿ Common Hornbl.	Mg	// 010
PARGASITE – FE-HASTINGSITE	Fe ⊢━┿━━ Fe-Hast. Hast.	Mg ━━┥ Pargasite	// 010
OXYHORNBLENDE	.⊢━━┥		// 010
BARKEVIKITE	⊢┥		// 010
GLAUCOPHANE – RIEBECKITE	Fe ⊢━━┿ Mg ⊣ Glauc. Mg ⊢━━┿━ Fe Riebeck.		Glauc. // 010 Rieb. ⊥ 010
ECKERMANNITE – ARFVEDSONITE	Eck. ⊢━━┿━┥ ⊢━━━┥ Arfv		Eck // 010 Arfv. ⊥ 010
KATAPHORITE	⊢━━━┥		Mg-Kat. // 010 Fe-Kat. ⊥ 010
	2V$_X$ 0° 20° 40° 60° 80° − ve	100° 120° 140° + ve	

Fig. 96. Ranges of 2V and orientation of the O.A.P. in the amphiboles.

Orientation diagrams and extinction angles. *Note*: cleavage fragments of any amphibole will lie parallel to the c-axis and one of the cleavages, and will always provide off-centred interference figures.

(1) *Orthorhombic amphiboles* (Fig. 97). Basal sections (Fig. 97a) have symmetrical extinction. All prismatic sections (Fig. 97b, c) have straight extinction with Z parallel to c. Maximum interference colours are shown by (010) sections which display the traces of the cleavages parallel to the c-axis (Fig. 97c). (100) and (001) sections provide bisectrix interference figures.

(2) *Monoclinic amphiboles with O.A.P.* // *(010)* (Fig. 98). Sections to which c is perpendicular have symmetrical extinction (Fig. 98a). Sections approximately parallel to (100) which display no cleavage (Fig. 98b) often provide good optic-axis interference figures, and display very little pleochroism. In (010) sections (Fig. 98c), it is possible to measure the extinction

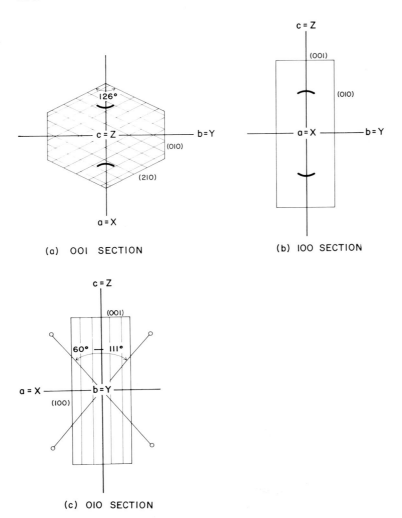

(a) OOI SECTION

(b) IOO SECTION

(c) OIO SECTION

Fig. 97. Orientation diagrams for the orthorhombic amphiboles.

angle $Z \wedge c$ (Table IV). It is important to choose the (010) section carefully; being the XZ plane, this section will display the highest interference colour for the particular amphibole.

(3) *Monoclinic amphiboles with O.A.P.* $\perp$ *(010)* (Fig. 99). It is customary to measure the extinction angle $X \wedge c$ (Table IV), although in practice this may be difficult because of the strong dispersion which results in poorly defined extinction.

Colour and pleochroism. The colours and pleochroic schemes for the amphi-

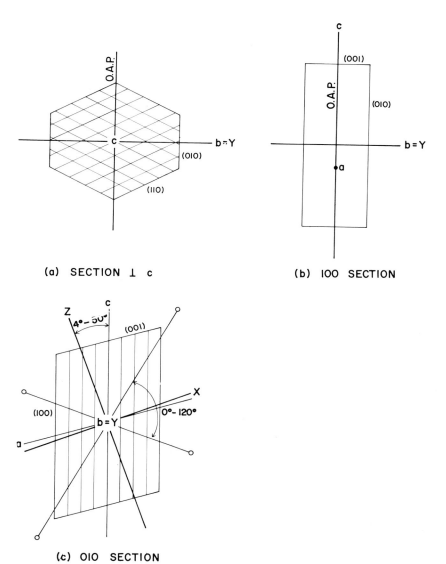

(a) SECTION ⊥ c

(b) 100 SECTION

(c) 010 SECTION

Fig. 98. Orientation diagrams for the monoclinic amphiboles with O.A.P. parallel to (010).

boles in thin section are summarised in Fig. 100. The pleochroic scheme is usually $Z > Y > X$, or $Z = Y > X$, or $Z > Y = X$, but is $X > Y > Z$ for eckermannite—arfvedsonite, and some riebeckite, and $Z < Y > X$ for Mg-kataphorite. The colours are darker in detrital grains.

TABLE IV

Extinction angles in the monoclinic amphiboles

Mineral	$Z \wedge c$ (Fig. 98c), O.A.P. // (010)	$X \wedge c$ (Fig. 99), O.A.P. $\perp$ (010)
Cummingtonite	15°(Fe)—21°(Mg)	
Grunerite	10°(Fe)—15°(Mg)	
Tremolite	15°(Fe)—21°(Mg)	
Actinolite	10°(Fe)—15°(Mg)	
Hornblendes	12°—34°	
Pargasite	20°—25°	
Hastingsite	12°(Fe)—20°(Mg)	
Oxyhornblende	4°—14°	
Barkevikite	11°—18°	
Glaucophane	4°—14°	
Eckermannite	30°—50°	
Mg-kataphorite	8°—16°	
Riebeckite		3°—21°
Arfvedsonite		0°—30°
Fe-kataphorite		36°—70°

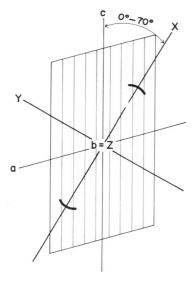

OIO SECTION

Fig. 99. Orientation diagram for monoclinic amphiboles with O.A.P. perpendicular to (010).

	COLOURLESS	YELLOW-BROWN Pale —— Dark	RED-BROWN Pale —— Dark	GREEN Pale —— Dark	BLUE-GREEN Pale —— Dark	BLUE Pale —— Dark
ANTHOPHYLLITE and GEDRITE	X Y Z	X Y- Z-		X Y- Z-	Y Z-	
CUMMINGTONITE — GRUNERITE	X Y Z (Mg-rich)	X Y- Z-		Z		
TREMOLITE — FE-ACTINOLITE	X Y Z (Mg-rich)	X Y-		X Y- Z-		
HORNBLENDE	X	X Y— Z —	Y— Z —	X Y— Z —	X Y— Z —	
OXYHORNBLENDE		X— Y— Z	Y— Z	X—		
BARKEVIKITE		X— Z	Y— Z			
GLAUCOPHANE	X	X		X	X	Y Z—
RIEBECKITE		Z		Z	Z	X— Y— Z —
ECKERMANNITE — ARFVEDSONITE		—Y —Z			—Y —Z	X
KATAPHORITE		—X—	—Y— —Z—	—Z—	—Y—	

Fig. 100. Pleochroic schemes and possible colours in the amphiboles.

Occurrence and habit

All amphiboles may occur as detrital grains in sedimentary rocks. Listed below are their occurrences in igneous and metamorphic rocks.

Anthophyllite and gedrite occur in a wide range of metamorphic rocks — often with cordierite in gneisses, and ascribed by some to metasomatism; they form with talc during the regional metamorphism of ultrabasic rocks. Sometimes grow as rims around pyroxene during retrograde metamorphism. Usually strongly prismatic or asbestiform.

Cummingtonite—grunerite occurs almost exclusively in metamorphic rocks. Cummingtonite may be present together with anthophyllite or hornblende in amphibolites; it rarely occurs in some volcanic rocks (dacites). Grunerite is found in metamorphosed Fe-rich sediments. Usually strongly prismatic, fibrous or asbestiform.

Tremolite—ferro-actinolite is the typical amphibole formed during greenschist facies or low-grade thermal metamorphism of basaltic or impure carbonate rocks. *Nephrite* is composed essentially of very fine-grained tremolite—actinolite. Some *uralite* is tremolite—actinolite formed by the replacement of pyroxene in basic igneous rocks. Usually strongly prismatic, fibrous or asbestiform.

Hornblende has a very widespread occurrence. It is the typical amphibole formed during amphibolite facies or high-grade thermal metamorphism of basaltic or impure carbonate rocks. Abundant in many plutonic and volcanic igneous rocks, especially diorites and granites. Hornblendes are usually Mg-rich in basic rocks, and Fe-rich in acid rocks. Some *uralite* is hornblende formed by the replacement of pyroxene in basic igneous rocks. Hornblende may form intergrowths with spinel in corona structures around olivine and pyroxene in basic igneous rocks, or intergrowths with plagioclase around garnet in eclogites. Usually stubby to long prismatic crystals.

Oxyhornblende occurs as phenocrysts in a wide variety of volcanic rocks. Usually short to moderately long prismatic crystals.

Barkevikite has a restricted occurrence as phenocrysts in alkaline igneous rocks. Usually short to moderately long prismatic crystals.

Glaucophane—riebeckite. Glaucophane is characteristic of the greenschist facies and the higher-pressure lawsonite—glaucophane facies of regional metamorphism. Riebeckite occurs in alkaline and acid plutonic and volcanic rocks, and also in metamorphosed Fe-rich rocks where it may be fibrous (crocidolite). Usually moderately elongate prismatic or fibrous.

Eckermannite—arfvedsonite has restricted occurrence in alkaline pluton-ic igneous rocks. Usually moderately elongate prismatic crystals.

Kataphorite is rare, and restricted to basic alkaline igneous rocks.

Alteration products. Tremolite and anthophyllite may alter to talc. Tremo-lite—ferro-actinolite is often altered to chlorite, and hornblende to chlorite or biotite, sometimes with iron oxides. The Fe-rich amphiboles are some-times replaced by iron oxides and siderite.

Distinguishing features of particular amphibole species

Anthophyllite and gedrite. Pale coloured or colourless, usually elongate or fibrous crystals, and restricted to metamorphic rocks (and as detrital grains). May be confused with cummingtonite—grunerite or tremolite—ferro-actino-lite. Distinguished as follows: straight extinction in (010) sections; generally lower δ than cummingtonite—grunerite; lack of twinning.

Cummingtonite—grunerite. Pale coloured or colourless, usually elongate or fibrous crystals, and usually in metamorphic rocks (or as detrital grains). May be confused with anthophyllite, gedrite, and tremolite—ferro-actinolite. Distinguished as follows: cummingtonite is +ve whereas tremolite is —ve; grunerite has a higher R.I. and δ than actinolite; inclined extinction, general-ly higher δ, and twinning distinguishes the series from anthophyllite and gedrite.

Tremolite—ferro-actinolite. Pale coloured or colourless, usually elongate or fibrous crystals, and restricted occurrences in metamorphic rocks, as an alteration product of pyroxene, or as detrital grains. May be confused with anthophyllite, gedrite, and cummingtonite—grunerite. Distinguished as fol-lows: tremolite is —ve whereas cummingtonite is +ve; actinolite has a lower

R.I. and δ then grunerite; inclined extinction and twinning distinguishes the series from anthophyllite and gedrite.

Hornblendes. Usually distinguished by their moderately strong green, blue-green, or brown colour, and lack of a strongly elongate or fibrous habit. They may be difficult to distinguish from strongly coloured ferro-actinolite or grunerite, though ferro-actinolite usually has a smaller extinction angle and grunerite has a higher δ. Particular types of hornblende are not generally easy to distinguish, but pargasite is characterised by its +ve nature, and ferro-hastingsite by its small −ve $2V$.

Oxyhornblende. In common with barkevikite and some kataphorite, oxy-hornblende is characterised by its intense brown and red-brown colour. Distinguished from barkevikite by its larger $2V$, larger δ, and generally smaller extinction angle, and from kataphorite by its higher R.I. and larger $2V$. Oxyhornblende only occurs as phenocrysts in volcanic rocks (and as detrital grains), but has a more widespread occurrence than barkevikite and kataphorite.

Barkevikite. Intense brown or red-brown colour in common with oxy-hornblende and some kataphorite. Distinguished from oxyhornblende by its smaller $2V$, smaller δ, and generally larger extinction angle, and from kataphorite by its higher R.I. Barkevikite only occurs as phenocrysts in alkaline volcanic rocks.

Glaucophane—riebeckite. The pale-blue colour of most glaucophane is diagnostic. Riebeckite is characterised by its deep blue colour and O.A.P. ⊥ (010). Riebeckite does not always extinguish completely. Mg-rich glauco-phane is restricted to metamorphic rocks (and as detrital grains).

Eckermannite—arfvedsonite. Distinguished by its pleochroic scheme. Arf-vedsonite rarely extinguishes completely. Arfvedsonite has its O.A.P. ⊥ (010) and a characteristic extinction angle. Restricted to plutonic alkaline rocks.

Kataphorite. Mg-kataphorite is distinguished from other red-brown amphi-boles by its smaller $2V$. Fe-kataphorite is distinguished by its O.A.P. ⊥ (010), its extinction angle, and pleochroic scheme. Restricted to basic alkaline igneous rocks.

Minerals likely to be confused with amphiboles

The general properties of the coloured amphiboles are quite distinctive, and confusion with other minerals is not likely. Red-brown biotite is some-times confused with red-brown amphibole (they often occur together), but the single cleavage, mottled extinction, and pseudo-uniaxial interference fig-ure of the biotite enables the distinction to be made. (100) sections of hornblende are sometimes mistaken for uniaxial tourmaline, but these sec-tions usually provide good biaxial interference figures.

The colourless amphiboles may be confused with zoisite, sillimanite or wollastonite. However, all three minerals lack the characteristic amphibole cleavages, zoisite often has anomalous interference colours, and wollastonite and some zoisite have their O.A.P. across the cleavage or length of the crystals. Fibrous anthophyllite (or gedrite) and sillimanite may be very difficult to distinguish, although their occurrences are usually different. Sillimanite has a slightly higher R.I. than the common range of anthophyllite and gedrite.

Fibrous amphiboles have a higher R.I. than fibrous serpentine.

E. TECTOSILICATES

In the tectosilicates, all four oxygen atoms of each $[SiO_4]$ tetrahedral group are shared with adjacent tetrahedra so that a framework of general formula $[SiO_2]$ is formed. The silica minerals (including quartz) are composed solely of such frameworks, and have the formula SiO_2. In the other tectosilicates, some of the Si is replaced by Al, and other elements such as Na, K and Ca balance the electrical charges in the formulae. Tectosilicates described here are: the silica minerals; the feldspars; the feldspathoids; the zeolites; and scapolite.

The silica minerals

SiO_2 occurs naturally as a number of polymorphs of which the most important and abundant is trigonal low-quartz. At atmospheric pressure, low-quartz rapidly transforms above $573°C$ to the closely similar hexagonal high-quartz. High-quartz always inverts to low-quartz below $573°C$. Above $870°$, high-quartz may transform to hexagonal tridymite, which in turn transforms above $1470°C$ to cubic cristobalite (Frondel, 1962). These transformations are sluggish and do not always take place, either with increasing or decreasing temperature. Hence, unstable low-forms of tridymite and cristobalite may occur at low temperatures. The structures of tridymite and cristobalite are very open, and they readily accept impurities, and it is found that in some cherts (Lancelot, 1973) and hydrothermal deposits (e.g. siliceous sinters), low forms of tridymite and cristobalite grow in the presence of impurities at low temperatures. At very high pressures (in excess of 20 kilobars) the polymorphs *coesite* and *stishovite* may form. These high-pressure polymorphs are found naturally in meteorite craters; they are not described here.

In addition to the above polymorphs, quartz may occur as the microcrystalline or fibrous form known as chalcedony; chalcedony differs from quartz in properties such as R.I. due to the admixture of varying quantities of water and other impurities. Opal is a form of hydrous silica which consists of cryptocrystalline low-cristobalite or, in the precious variety, regularly packed spheres of amorphous silica.

No. 68. QUARTZ (+ve)   [SiO$_2$]

Trigonal. No good cleavage. In volcanic rocks, quartz phenocrysts are euhedral bipyramidal hexagonal prisms (Fig. 101a) — originally hexagonal high-quartz; in quartz veins, crystals may be euhedral singly-terminated pseudohexagonal elongate prisms; quartz is commonly anhedral, filling the gaps between other crystals in igneous rocks or forming mosaics in metamorphic rocks (Fig. 101b). Parallel twinning on the Dauphiné (twin axis = c-axis) and Brazil (twin plane = $\{11\bar{2}0\}$) laws very common, but cannot be observed in thin sections, since the c-axes of the twins are parallel (Frondel, 1962).

Colour in thin section: colourless.

Colour in detrital grains: usually colourless, but may be weakly coloured in a wide variety of tints.

Optical properties: uniaxial +ve (rarely anomalously biaxial).
ω = 1.544, ϵ = 1.553.
δ = 0.009.
Euhedral prismatic crystals have straight extinction, and are length slow. Bipyramidal crystals in volcanic rocks have symmetrical extinction.

Deformation and recrystallisation. Quartz deforms easily by translation gliding and Dauphiné twinning (Tullis et al., 1973), especially in the presence of water which weakens the bonding in the SiO$_2$ structure. Translation gliding may take place on several systems, but most importantly parallel to the base $\{0001\}$ and the prisms, and less importantly parallel to the rhombohedra. Translation allows bending, slippage, and rotation within the quartz grains. Evidence that translation has taken place may be provided by *deformation lamellae*, which are narrow, closely spaced, subplanar features with a slightly different R.I. from the host quartz. Frequently, however, such evidence is destroyed during subsequent recrystallisation, which involves the nucleation of relatively strain-free grains and

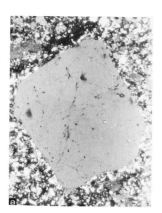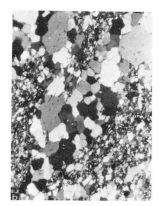

Fig. 101. Quartz: (a) typical section of bipyramidal quartz in dacite (view measures 1.7 mm × 1.2 mm); (b) mosaic of recrystallised quartz grains in biotite schist, Taipo Valley, New Zealand (view measures 1.8 mm × 1.3 mm); (c) deformation bands and undulatory extinction, Greenland Group, New Zealand (view measures 1.8 mm × 1.3 mm). Crossed-polarised light.

the growth of these at the expense of the deformed quartz; such recrystallisation may result in the development of a mosaic characterised by strain-free grains with subplanar boundaries and triple-point junctions (Fig. 101b). *Deformation bands* and *undulatory extinction* are very commonly observed in quartz (Fig. 101c), and probably result from a process of polygonisation, that is the migration of dislocations (resulting from translation) so that they are concentrated in zones perpendicular to translation planes; in other words, they represent a partial recovery following deformation. Since basal slip is the most common translation system, deformation bands are most commonly parallel to the *c*-axis of quartz. Deformation and recrystallisation often result in the development of a preferred orientation of the *c*-axes of quartz (Shelley, 1971, 1972; Tullis et al., 1973); a quantitative assessment of this is made using the U-stage or X-rays, but a qualitative assessment can be made in thin section by using the sensitive-tint plate, and observing the preferred orientation of the fast directions (= ω, $\perp$ *c*-axis) in a quartz mosaic.

Occurrence. Quartz is highly resistant to weathering, and is an abundant detrital mineral being the prime constituent of many sandstones. In addition, the secondary diagenetic formation of quartz is common. Many cherts are made primarily of quartz. It is stable throughout the entire range of metamorphism, and is important in a wide variety of metamorphic rocks; it is the principal constituent of quartzite. It is found in a wide variety of igneous rocks, and is an essential component of some such as granite and rhyolite. Phenocrysts of quartz in volcanic rocks are usually bipyramidal, and originated as the high-temperature hexagonal form; such phenocrysts are commonly corroded. Quartz is often intergrown with feldspar in igneous rocks to form graphic and granophyric intergrowths and myrmekite (see under feldspars). It does not occur in igneous rocks containing feldspathoids. Quartz is also an important hydrothermal mineral occurring in veins.

Distinguishing features. Quartz is distinguished by its lack of colour, cleavage and visible twinning, by its low relief and δ, by its uniaxial +ve character, and by the lack of alteration. It is most commonly confused with the feldspars or cordierite. Both these minerals are commonly twinned or altered and in the case of feldspar, cleaved, but in the absence of these features, distinction can be difficult. The R.I. of cordierite and feldspar may be distinctive, but Ca-oligoclase, Na-andesine, and some cordierite have the same R.I. as quartz. The uniaxial +ve character of quartz is diagnostic, but if quartz and feldspar or cordierite occur together, it may be necessary to use U-stage (Chapter 4) or staining techniques (see under feldspars and cordierite) to quantitatively assess the *amount* of quartz present. Less commonly, beryl, scapolite, and nepheline may be confused with quartz, but all these minerals are uniaxial —ve, and beryl, nepheline, and most scapolite differ in their R.I.

No. 69. CHALCEDONY (+ve)   [SiO_2] + <10% H_2O

Essentially microcrystalline quartz with loosely held water, but there may be some substitution of OH for O in the [SiO_4] tetrahedra.

Crystals are of quartz; microcrystalline and often fibrous, the fibres being elongate parallel to [$11\bar{2}0$], or less commonly [$10\bar{1}0$] or the *c*-axis [0001].

Colour in thin section: colourless.

Optical properties: uniaxial +ve, but aggregates often give biaxial +ve interference figures. ω = 1.526—ca. 1.544, ϵ = 1.532 (?)—ca. 1.553. δ = 0.005—ca. 0.009.

Optical properties may be difficult to observe due to fine grain size. Fibres are fast or, less commonly, slow along.

Occurrence. Chalcedony occurs filling vesicles and cavities in igneous rocks, in veins and sinter deposits, in cherts, and replacing a variety of rocks, particularly volcanics and those rich in carbonates. Well-known varieties of chalcedony include *agate* (banded), *jasper* (red), *flint* (massive, nodular) and *chert.* Chalcedony often grades into quartz within a specimen.

Distinguishing features. The occurrence, fibrous or microcrystalline nature, low relief and δ, and lack of alteration are distinctive. Microcrystalline chalcedony is very similar to the felsitic groundmass of acid volcanic rocks. A felsitic groundmass is composed, however, of alkali-feldspar and quartz, and the presence of these two minerals is easily detected by closing the sub-stage diaphragm and observing Becke lines.

No. 70. TRIDYMITE (+ve) $\qquad$ [SiO_2] + Al, Na, etc. impurities?

High-temperature tridymite is hexagonal, low-temperature tridymite orthorhombic. No good cleavage. Crystals are usually very small hexagonal or pseudo-hexagonal {0001} plates. Wedge-shaped or sector twins common.

Colour in thin section: colourless.

Optical properties: usually biaxial +ve. $2V_z = 30°-90°$.
$\alpha = 1.468-1.479$, $\beta = 1.469-1.480$, $\gamma = 1.473-1.483$.
$\delta = 0.002-0.007$ (usually <0.005). $Z = c$.
Plate-like crystals are fast along.

Occurrence. In cavities, vesicles, the groundmass, or less commonly as phenocrysts in volcanic rocks, especially rhyolites, trachytes, and andesites. Also occurs in siliceous sinters and high-grade thermally metamorphosed sandstones. Some cherts contain tridymite (Oehler, 1973), and some workers (Lancelot, 1973) believe such deposits to form instead of quartz-cherts where impurities contributed by clays are readily available.

Distinguishing features. The low R.I. and δ, the small plate-like crystals with twinning, and occurrence are distinctive. Tridymite has a slightly lower R.I. than cristobalite. Some of the zeolites (e.g. chabazite and heulandite) are similar, but usually have a higher R.I. and occur in association with basic igneous rocks.

No. 71. CRISTOBALITE (−ve) $\qquad$ [SiO_2] + Al, Na, etc. impurities?

High-temperature cristobalite is cubic; low-temperature cristobalite is tetragonal. No cleavage. Euhedral crystals are usually octahedra; also spherulitic; in cherts, cristobalite may form spherules consisting of tridymite-like plates. Penetration twins may result during inversion from the high to low form.

Colour in thin section: colourless.

Optical properties: uniaxial −ve or pseudo-isotropic.
$\omega = 1.487$, $\epsilon = 1.484$.
$\delta = $ ca. 0.003.

Occurrence. In cavities, vesicles, and the groundmass of volcanic rocks, especially rhyolites, trachytes and andesites. Also found in siliceous sinters. In the presence of impurities contributed by clay minerals, cherts may develop cristobalite and tridymite (Lancelot, 1973, Oehler, 1973) instead of quartz at low temperatures.

Distinguishing features. The low R.I., the very low δ and occurrence are distinctive. The R.I. is slightly higher than that of tridymite. Some zeolites (e.g. chabazite) are similar, but they generally occur in basic volcanics.

No. 72. OPAL (isotropic) $[SiO_2]$ + $<20\%$ H_2O

Cryptocrystalline low-cristobalite, or amorphous (precious varieties) (Jones et al., 1964).

Colour in thin section: usually colourless, but may be pale coloured, especially brown.

Optical properties: isotropic.
n = ca. 1.41—1.47, usually ca. 1.435—1.455.
R.I. increases with decrease in H_2O.

Occurrence. A secondary mineral in veins and cavities in igneous rocks, in sediments neighbouring areas of volcanic activity, and in siliceous sinters. May be derived from organic remains, and the prime constituent of diatomite and some cherts. Also in other sediments, often as concretions or replacing fossils, and usually derived from siliceous organic remains, or as a result of weathering under arid conditions. Opal may grade into chalcedony and quartz.

Distinguishing features. The low R.I. (moderate relief), isotropic character, and lack of crystal shape and cleavage are distinctive.

The feldspars

Feldspars are the most common rock-forming minerals in the earth's crust. They are extremely abundant in most igneous rocks, and the presence or absence of particular feldspar species is the basis of igneous rock classification. Feldspars are also important components of many metamorphic rocks and are common in some sediments, especially sandstones.

The feldspars are divided into two main solid solution series: (1) the alkali-feldspars, a series from $KAlSi_3O_8$ (Or) to $NaAlSi_3O_8$ (Ab); and (2) the plagioclase-feldspars, a series from $NaAlSi_3O_8$ (Ab) to $CaAl_2Si_2O_8$ (An). There is no solid-solution series between $KAlSi_3O_8$ and $CaAl_2Si_2O_8$. Rare Ba-feldspars (*hyalophane* and *celsian*) exist but will not be considered further.

The alkali- and plagioclase-feldspar series are subdivided further according to the particular proportions of the molecules Or, Ab and An, and according to their structural state. The structural state is a reflection of thermal history, and allows feldspars to be subdivided into "high-" or "low-temperature" types. "High-temperature" feldspars are those which crystallised at high temperatures and cooled sufficiently rapidly (e.g. in some volcanic rocks) to preserve the structure and optics of high temperature forms. "Low-temperature" feldspars either result from slow cooling of high-temperature forms (e.g. in some plutonic rocks) or crystallise initially at low temperatures (e.g. in some metamorphic rocks). Feldspars may also have "intermediate" structures and optics.

No. 73. ALKALI-FELDSPAR (—ve, also +ve)   (K,Na)[AlSi₃O₈]

includes: No. 73A — microcline (—ve), triclinic; No. 73B — orthoclase (—ve), mono-clinic; No. 73C — sanidine (—ve), monoclinic; No. 73D — anorthoclase (—ve), triclinic; No. 73E — adularia (—ve), monoclinic or triclinic; No. 73F – pericline (+ve), triclinic; and albite, normally discussed under plagioclase (No. 74A).

The terminology based on composition and structural state is explained in Fig. 102. The boundaries between particular species are based partly on crystal symmetry, but are otherwise somewhat arbitrary. The solvus curve represents the limits of solid solution.

At low pressures and high temperatures (Fig. 102a), there is complete solid solution between Or and Ab. The high-temperature forms may be metastably preserved by quick cooling. Slow cooling results in inversion to low-temperature forms, and may result in unmixing and the separation of a K-rich from a Na-rich feldspar if the solvus curve is intersected; this unmixing or exsolution is often represented by an intergrowth of the two feldspars known as "perthite" (Fig. 120) "mesoperthite" or "antiperthite". At higher pressures (Fig. 102b), less solid solution is possible, and melts of inter-mediate composition will crystallise two separate feldspars from the start;

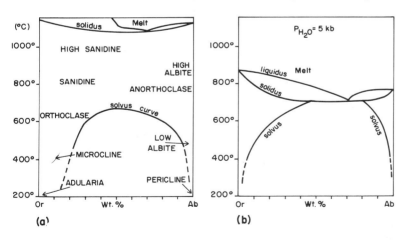

Fig. 102. (a) Terminology of the alkali-feldspars and positions of solidus and solvus curves at low pressures (curves redrawn with permission from Tuttle and Bowen, 1958). (b) Solidus and solvus curves at P_{H_2O} = 5 kbars. Redrawn with permission from Morse (1970).

TABLE V

Subdivision of plagioclase-feldspar (mol.%)

	Ab (%)	An (%)
No. 74A. Albite	100—90	0— 10
No. 74B. Oligoclase	90—70	10— 30
No. 74C. Andesine	70—50	30— 50
No. 74D. Labradorite	50—30	50— 70
No. 74E. Bytownite	30—10	70— 90
No. 74F. Anorthite	10— 0	90—100

unmixing takes place in the same way as at low pressures if the solvus curve is intersected during slow cooling.

Adularia and pericline crystallise in veins from hydrothermal solutions, and usually have compositions closely approaching the two pure end-members of the series.

No. 74. PLAGIOCLASE-FELDSPAR (+ve and —ve)

A series from $Na[AlSi_3O_8]$ to $Ca[Al_2Si_2O_8]$ subdivided according to the percentage proportions of the two molecules Ab and An (Table V).

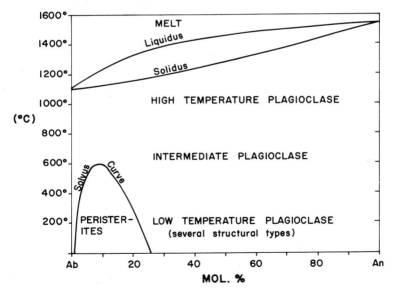

Fig. 103. Terminology, liquidus, solidus and peristerite solvus curves for the plagioclase feldspars. Based, with permission, on information from Bowen (1913) and Barth (1969).

The plagioclase-feldspars are qualified as "high-temperature", "inter-mediate", or "low-temperature" (Fig. 103). Complete solid solution at high temperatures breaks down at lower temperatures in a complicated and not fully understood way, but the most important feature is the peristerite solvus curve (Fig. 103). For igneous rocks which cool slowly and contain oligoclase, this curve represents the unmixing of the oligoclase into a more sodic and a more calcic plagioclase which together form an intergrowth called peristerite; this intergrowth is too fine to be seen with the microscope, and in optical mineralogy the peristerite may be called low-temperature oligoclase. For metamorphic rocks in which feldspars first crystallise at low temperatures, the curve prescribes those compositions that are possible; hence during progressive metamorphism, early pure albite does not become progressively more calcic, but is succeeded abruptly by calcic oligoclase or andesine.

The feldspars are so important in petrology that it is necessary to specify the type as precisely as possible, and to do this a number of special methods have been devised. To employ these, the student must first understand the crystallography and twin-laws of the feldspars.

Crystallography and twin-laws

Feldspars belong to either the monoclinic (high-sanidine, sanidine, ortho-clase, and some adularia) or the triclinic crystal systems. There is little varia-tion in the axial ratios and the crystallographic angles among the feldspars. For example: monoclinic orthoclase has an axial ratio $a : b : c = 0.658 : 1 : 0.555$, and $\alpha = 90°$, $\beta = 116°00'$, $\gamma = 90°$; triclinic anorthite has an axial ratio $a : b : c = 0.635 : 1 : 0.557$, and $\alpha = 93°10'$, $\beta = 115°50'$, $\gamma = 91°15'$.

The crystal habits of all the feldspars are similar, varying from stubby prismatic to tabular (with (010) dominating) or lath-like (usually elongate parallel to a) (Fig. 104). All feldspars have perfect $\{001\}$ and distinct $\{010\}$ cleavages.

Twinning is very common in all feldspars. The more important twin-laws are summarised in Table VI. For details of the other less common twin-laws

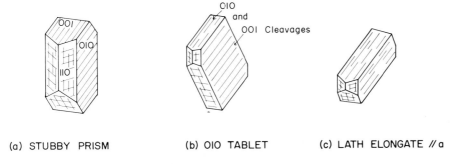

(a) STUBBY PRISM   (b) OIO TABLET   (c) LATH ELONGATE // a

Fig. 104. Common habits of feldspar.

TABLE VI

Common twin-laws of the feldspars

Name	Twin axis	Composition plane	Remarks
Normal twins			
Albite	$\perp$ (010)	(010)	often repeated*; very common; only in triclinic feldspar
Manebach	$\perp$ (001)	(001)	usually two individuals; not common
Baveno	$\perp$ (021)	(021)	usually two individuals; not common
Parallel twins			
Carlsbad	c	(010) or (100) or irregular	usually two individuals; very common
Pericline	b	rhombic section (see text)	often repeated*; very common; only in triclinic feldspar
Acline	b	(001)	often repeated*, not common; only in triclinic feldspar
Complex twins			
Albite-Carlsbad	$\perp$ c in (010)	(010)	usually a small number of individuals; common; only in triclinic feldspar

* Repeated twinning is also called multiple or polysynthetic twinning. Feldspar crystals may display any combination of twin-laws. Hence plagioclase may display within one crystal display Carlsbad, albite, and pericline twinning.

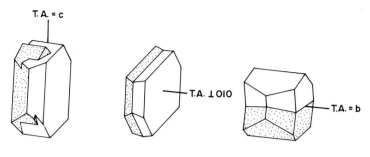

(a) PENETRATION   **(b) ALBITE-TWIN**   **(c) PERICLINE TWIN**
CARLSBAD-TWIN

Fig. 105. Common twin relationships in feldspar.

see Barth (1969) or Slemmons (1962). Fig. 105 illustrates some of the more common twin relationships.

The rhombic section (the composition plane for pericline twinning) is defined as the section that contains b, and intersects (010) in a line which is perpendicular to b (Tunell, 1952). The position of the section is very sensitive to changes in the value of the crystallographic angle γ. For example, the rhombic section is close to (10$\bar{1}$) in microcline, but close to (001) for anorthoclase (Fig. 106). In the plagioclase feldspars, the angle between a and the trace of the rhombic section on (010) varies both with composition and structural state (Fig. 107).

Origin of twinning in feldspar. Feldspars may twin during growth (primary or growth twinning), during subsequent deformation (secondary deformation twinning), or during inversion from monoclinic to triclinic symmetry (secondary transformation twinning) (Vance, 1961, 1969). Carlsbad, Manebach, Baveno and albite-Carlsbad twins can only form during growth. Albite and pericline twins can be of growth, deformation or transformation origin.

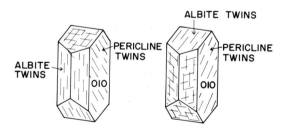

(a) MICROCLINE   **(b) ANORTHOCLASE**

Fig. 106. Orientation of albite and pericline twins in (a) microcline and (b) anorthoclase.

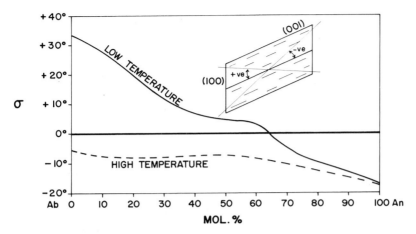

Fig. 107. Angle σ between a and the trace of the pericline twin-plane on (010) for high- and low-temperature plagioclase. Redrawn with permission from Starkey (1967).

Growth twins are often made of only two individuals, but if the twins are repeated, the spacing is irregular (Fig. 108a). Growth twins may be nucleated during growth, but may also result from synneusis or the colliding of crystals in a melt to join together in a twinned orientation (Fig. 108b). Deformation twins are usually repeated and regularly spaced, and may be wedge-shaped if the crystal is bent (Fig. 108c). Transformation twins (on the albite and pericline twin-laws) form in response to a change from monoclinic to tri-clinic symmetry in alkali-feldspar. The change commences at numerous sites in the crystal, and may be achieved by forming either of the albite or pericline twin orientations. The particular twin orientations first formed, spread and coalesce as the change of symmetry spreads through the crystal. The result is a characteristic pattern of interpenetrating albite and pericline twins called "cross-hatch" twinning (Fig. 108d). Transformation twinning is characteristic of microcline and anorthoclase which have formed by inver-sion from orthoclase and high-sanidine respectively.

Carlsbad twins are characteristic of magmatic rocks, and are very rare in metamorphic rocks. The reason for this is not clear since primary albite twins are common in metamorphic rocks. It may be that a high degree of supersaturation is necessary to nucleate Carlsbad twins, and that this is possi-ble only during the initial growth stage in magmas; this may explain why only one central Carlsbad composition plane is formed in most crystals.

Identification of the feldspars

The feldspars are distinguished from other minerals by the following gen-eral properties: R.I. in the range 1.518—1.590; δ in the range 0.005–0.013; $2V_x$ in the range $0°—105°$; two cleavages $\{001\}$ and $\{010\}$. Some of the

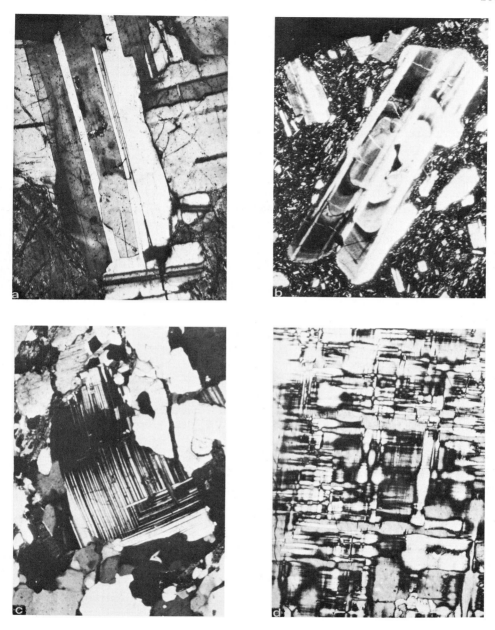

Fig. 108. (a) Growth twins in plagioclase, gabbro, Onawe, New Zealand (view measures 0.8 mm × 0.6 mm). (b) Twinning resulting from synneusis of two plagioclase crystals (note re-entrant angles), Te Pua Andesite, New Zealand (view measures 1.7 mm × 1.3 mm). (c) Secondary deformation albite and pericline twinning in plagioclase (note bending and wedging out of twin-lamellae), Constant Gneiss, New Zealand (view measures 1.8 mm × 1.3 mm). (d) "Cross-hatch" transformation twinning in microcline, granite, Bavaria (view measures 0.7 mm × 0.4 mm). Crossed-polarised light.

zeolites, particularly scolecite and laumontite, have similar properties to the low R.I. feldspars, but have quite different crystal habits, occurrence and orientation diagrams.

The procedures for identifying specific feldspars are described separately for thin sections and crushed-grain mounts. Staining methods are also described briefly.

Thin sections. The procedure is as follows.

General approach

The first step is to estimate R.I. (Fig. 109). This can be done *qualitatively* in thin section by comparison with the R.I. of canada balsam and/or quartz (if present confirm by obtaining a uniaxial +ve interference figure), and by observing relief and Becke lines (Table VII). In fact, a quite accurate estimation of sodic plagioclase composition can be made by comparing α (fast) and γ (slow) in the highest interference colour grains with canada balsam, or with ω and ϵ of quartz (Fig. 109). Note that the fast direction of any quartz grain is ω, and that grains with the highest interference colour contain ω and ϵ.

Having thus *estimated* the composition of the feldspar, more specific identifications are made as follows.

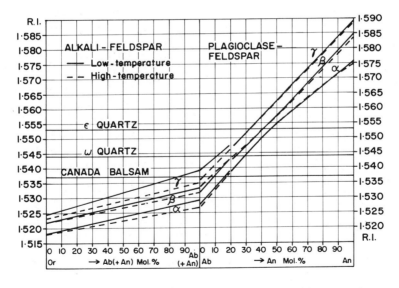

Fig. 109. R.I. variation with feldspar composition. Based with permission on curves in Tröger (1971) and Smith (1958, fig. 3).

TABLE VII

Estimation of feldspar composition from refractive index

Feldspar	R.I. compared with C.B. and quartz	Becke line	Relief
Anorthite, bytownite and labradorite	distinctly above	strong	moderate
Andesine	just above quartz, distinctly above C.B.	strong	low
Calcic oligoclase	same as quartz, just above C.B.	weak	low
Sodic oligoclase	just below quartz, same as C.B.	very weak	low
Albite	distinctly below quartz, just below C.B.	weak	low
K-feldspar	distinctly below	strong	low

Note: The lower diaphragm of the microscope must be at least partly closed to observe the Becke lines.

Alkali-feldspar

To *accurately* determine the composition, a precise R.I. measurement of crushed grains must be made; alternatively, chemical or X-ray methods can be used (Bowen and Tuttle, 1950; Fraser and Downie, 1964). However, the main types can be *roughly* determined using $2V$ and twinning characteristics. $2V$ varies with structural state and composition as shown in Fig. 110.

For K-rich feldspars, a small $2V$ indicates sanidine, a moderate $2V$ orthoclase, and a high $2V$ microcline. K-rich high-sanidine has the O.A.P. parallel to (010), whereas the other alkali-feldspars have their O.A.P. perpendicular to (010) (Fig. 111). Albite and pericline twinning are impossible in orthoclase and sanidine, but microcline and anorthoclase are triclinic and normally display the transformation-type cross-hatched twinning. They are distinguished from each other by occurrence (anorthoclase in volcanics, microcline in granites and metamorphic rocks) and also by the orientation of the pericline twin plane (Fig. 106); hence the cross-hatched effect is seen in the high interference colour sections (001) of microcline, but in the low interference colour sections (100) of anorthoclase.

Adularia is characterised by its occurrence in hydrothermal veins. It often has variable optics (from sanidine to microcline type) within a single crystal, and may display cross-hatched twinning.

The sodic alkali-feldspars are best treated as plagioclase. Albite is distinguished from the other alkali-feldspars by its higher R.I. Almost all albite

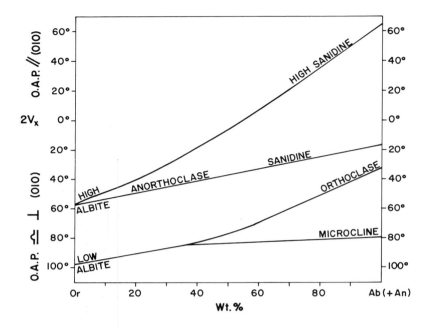

Fig. 110. $2V$ variation with composition and structural state of alkali-feldspar. Redrawn, with permission, from Tuttle (1952), and MacKenzie and Smith (1956, fig. 1).

crystallises initially in the triclinic system, so that the albite and pericline twinning, if present, is *not* of the transformation type. Low-albite is further characterised by being biaxial +ve. Pericline is found in hydrothermal veins as crystals elongate parallel to *b*.

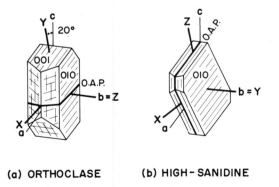

(a) ORTHOCLASE   (b) HIGH-SANIDINE

Fig. 111. Optical orientations of the alkali-feldspars.

Plagioclase-feldspar[*]

The plagioclase-feldspars are all triclinic, and the positions of X, Y and Z vary considerably with composition (Fig. 112) and structural state. $2V$ also varies with composition and structural state (Fig. 113), and is not a useful property by itself. However, these two variable properties taken together can be used to determine both composition and structural state. Proceed as follows:

(1) Estimate composition by measuring the extinction angles according to one of the three methods described below. These methods make use of the wide variation in the X, Y and Z positions. Two curves are provided for each method; the high-temperature curve should be used for volcanic plagioclase, and the low-temperature curve for plutonic and metamorphic plagioclase. Detrital plagioclase may be of either type.

(2) The choice of curve should then be checked by measuring $2V$ and referring to Fig. 113, using the composition estimated by the extinction-angle method. The estimation of composition should be judiciously modified to ensure that the same structural state is indicated by both $2V$ and extinction-angle values.

Cautionary notes. For strongly zoned plagioclase, it is generally necessary to use method 1, which enables the range of composition to be measured. Methods 2 and 3 assume that all plagioclase in one rock has the same composition; apart from the presence of zoning, this is normally the case for igneous and metamorphic rocks, but is an unjustified assumption for detrital plagioclase. Microlites in volcanics are usually elongate parallel to a, and for these, method 3 is the best.

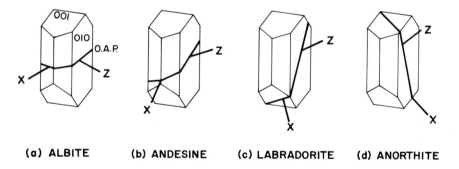

(a) ALBITE   (b) ANDESINE   (c) LABRADORITE   (d) ANORTHITE

Fig. 112. Variations in positions of X, Z, and the O.A.P. in plagioclase.

---

[*] Composition and the twin-laws of plagioclase crystals can be most accurately determined using the universal-stage methods described by Slemmons (1962) and Von Burri et al. (1967).

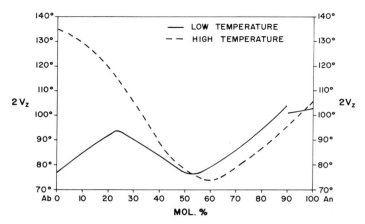

Fig. 113. Variation of $2V$ with composition and structural state of the plagioclase feld-spars. Redrawn with permission from Smith (1958, fig. 2).

Method 1. Extinction angle in section $\perp a$. The section $\perp a$ is recognised by the presence of two cleavages almost at right angles to each other (Fig. 114). Both cleavages should appear sharp (i.e. vertical) in thin section. Albite twins will often be visible parallel to the (010) cleavage. The extinction angle between the fast ray (α') and (010) is measured (it will always be less than $45°$), and the appropriate curve of Fig. 114 referred to. The convention for +ve and —ve angles is shown in Fig. 114, but the ambiguity for readings less than $20°$ can also be resolved by R.I., since plagioclase more sodic than An_{20} has $\alpha <$ canada balsam.

This method is the simplest and least ambiguous of the three, and is particularly appropriate for strongly zoned plagioclase (range of composition equals the variation in extinction angle within one crystal). Unfortunately, it is not always possible to find a suitably oriented section.

Method 2. The Michel-Lévy method. The *maximum* extinction angle of the fast ray (α') onto (010) is measured in sections normal to (010). Sections normal to (010) are recognised by the presence of sharply bounded albite twins parallel to the (010) cleavage. Albite twins are recognised by having even-illumination when the twin planes are oriented E—W, N—S, *and* in the $45°$ position (Fig. 115). Each twin orientation in turn is put into extinction, the fast direction identified, and the angle between the fast direction (α') and (010) measured. The two readings thus obtained are averaged. If the readings are not within $5°$ of each other they should be discarded. Readings must be made from at least seven suitably twinned crystals. The *maximum* reading obtained is selected, and the composition determined from Fig. 116. The ambiguity for readings less than $20°$ can be resolved by R.I., since plagioclase more sodic than An_{20} has $\alpha <$ canada balsam.

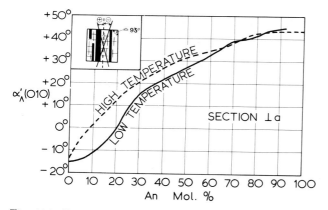

Fig. 114. Extinction angles in plagioclase sections perpendicular to *a*. Curves are based on the data of Von Burri et al. (1967, p. 307).

The method is not satisfactory for strongly zoned plagioclase, or detrital plagioclase, since the method assumes a restricted range of composition within the rock.

Method 3. Using combined albite and Carlsbad twins. The extinction angle of ($α'$) onto (010) is measured on each side of a Carlsbad twin. The presence of sharply bounded albite twins on each side of the Carlsbad twin enables the position of (010) to be accurately judged (Fig. 117). The albite twins are recognised by their even-illumination in the N—S, E—W, and 45° positions (Fig. 115). The Carlsbad twin does not have even-illumination in these positions, and remains visible, particularly in the 45° position.

Measure and average the readings of the fast direction onto (010) from the albite twins (as in the Michel-Lévy method) for each side of the Carlsbad twin. Two (usually different) values are thus obtained, and the composition determined from Fig. 118, using the solid curves for the larger angle and the dashed curves for the smaller angle. Positive and negative angles can only be determined in sections ⊥ *a* (Tobi, 1963), but if such a section is found, the

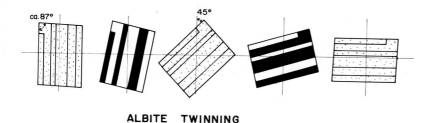

ALBITE TWINNING

Fig. 115. Recognition of albite twinning by the even-illumination in three positions.

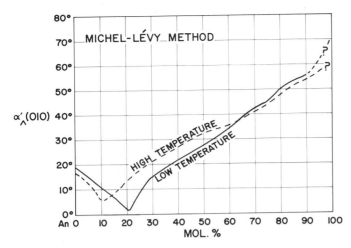

Fig. 116. Michel-Lévy extinction-angle method for plagioclase. Curves are derived from Fig. 118.

albite-Carlsbad method is superfluous and method 1 should be used. This means that readings greater than 20° are nearly always ambiguous, and for readings less than 20° there may be three possible solutions. Some ambigu-

Fig. 117. Combined albite and Carlsbad twinning in plagioclase. Bluff Complex, New Zealand (view measures 1.0 mm × 0.66 mm). Crossed-polarised light.

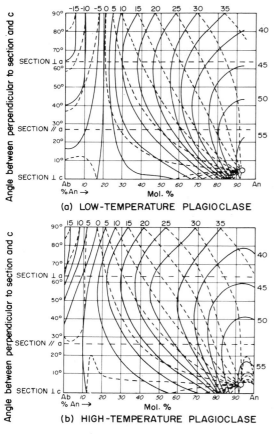

Fig. 118. Extinction-angle curves for combined albite and Carlsbad twinning in plagio-clase. Redrawn with permission from Tobi and Kroll (manuscript in preparation). See text for explanation.

ities may be resolved by R.I., but contrary to the explanations in other texts, it is often necessary to make at least two determinations to resolve them.

Because of the ambiguities, the method is less suitable for strongly zoned crystals, and not at all suitable for detrital plagioclase. Microlites in volcanics are normally elongate parallel to a, and often Carlsbad-twinned; this method is the best for determining their composition — use extinction values close to the horizontal line marked "section parallel to a" in Fig. 118.

Crushed-grain mounts. An estimation of the composition of alkali-feldspar in crushed grains can be made from $2V$ and twinning characteristics, as described for thin sections. A more precise determination is possible by accurately measuring R.I. (Fig. 109), using $2V$ to determine the structural state (Fig. 110).

The composition of crushed grains of plagioclase can also be determined by accurately measuring R.I. (Fig. 109), using $2V$ to specify the structural state (Fig. 113). However, there may be difficulties in finding suitably orientated grains to measure α, β or γ precisely, since the majority of crushed fragments will lie parallel to the $\{001\}$ or less commonly, the $\{010\}$ cleavages. An alternative to measuring R.I. precisely is to measure the extinction angles of α' onto the $\{010\}$ cleavage trace in fragments lying on (001), or of α' onto the $\{001\}$ cleavage trace in fragments lying on (010) (Fig. 119). $2V$ must be measured to determine the structural state, and provides a check on the choice of curve in Fig. 119, and hence allows an improvement in the accuracy of the determination. The ambiguous positive and negative angle values of Fig. 119 are sorted most easily on the basis of R.I.

Staining methods. Many rocks contain two or three of the minerals quartz, alkali-feldspar and plagioclase. For the purposes of modal analysis, it is necessary to identify every grain in a thin section (or rock slab) quickly, and this is not always easy with standard microscope procedures. To assist such analyses, various staining methods have been devised. The standard technique is to etch the specimen with HF, and stain K-feldspar yellow with sodium cobaltinitrite and plagioclase red by treating with $BaCl_2$ and K-rhodizonate or other dyes (Bailey and Stevens, 1960; Laniz et al., 1964). Lyons (1971) has described a more suitable procedure for rock-slab surfaces.

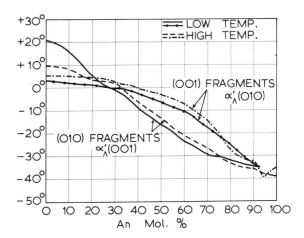

Fig. 119. Extinction angles for (010) and (001) cleavage fragments. Curves are based on the data of Von Burri et al. (1967, p. 307).

Occurrence and alteration

The alkali-feldspars are essential constituents of many alkaline and acid igneous rocks, e.g. granites, syenites, nepheline-syenites, and their volcanic equivalents. In plutonic rocks, the alkali-feldspar is usually orthoclase or microcline, but in volcanics, the higher-temperature forms sanidine and anorthoclase are common. Orthoclase and microcline also occur in pegmatites, often in graphic intergrowth with quartz (see below). Orthoclase and microcline are formed during regional and thermal metamorphism, orthoclase being particularly characteristic of high-grade pelitic rocks. Plagioclase-feldspars are essential constituents of a wide range of igneous rocks. The more calcic plagioclases are typical of basic rocks, the more sodic of acid and alkaline types. Plagioclase is also common in metamorphic rocks. The lower grades of metamorphism are characterised by almost pure albite, and there is a sudden jump in composition (Fig. 103 and accompanying text) to calcic oligoclase or andesine in the higher grades; this change coincides with the boundary between the greenschist and amphibolite facies of regional metamorphism. Calcic plagioclase is often abundant in thermally metamorphosed carbonate rocks. Plagioclase is unstable in the eclogite facies of metamorphism.

Adularia and pericline occur in hydrothermal veins, and both alkali- and plagioclase-feldspar (usually pure K and Na end members) may form during diagenesis.

Feldspars are prone to extensive weathering or alteration. The principal deuteric or hydrothermal alteration products are: sericite, composed of mica or illite — often fine-grained; saussurite (from plagioclase) — a mixture of zoisite, epidote, albite, chlorite, carbonates, and other minerals; clay minerals. Feldspars also weather readily to clay minerals, giving crystals a cloudy appearance. However, both alkali- and plagioclase-feldspars are abundant as detrital grains in some sediments, particularly sandstones.

Common feldspar textures and intergrowths

Since any work with the feldspars must necessarily involve the very common textures and intergrowths, a brief description of these is given here. No attempt is made to provide an exhaustive account or to provide any more than brief suggestions as to their origin.

Perthite and antiperthite. Perthite consists of discrete areas of sodic plagioclase within a host K-feldspar crystal. In antiperthite, the host is plagioclase. The term mesoperthite is used if the two feldspars are in more or less equal proportions. The enclosed feldspar may occur as lensoid areas variously described as rods, beads or strings (Fig. 120), or as irregular "flame-like" or blocky patches. Perthitic structures should not be confused with twinning from which they are easily distinguished by observing the contrast in R.I. using Becke lines. Rod, bead, string perthite and antiperthite are normally considered to be the result of exsolution as prescribed by the solvus curve of

Fig. 120. Perthite: sodic plagioclase (dark areas) in host K-rich feldspar; note cleavages run continuously from K-rich through Na-rich areas (view measures 3.3 mm × 2.1 mm). Crossed-polarised light.

Fig. 102. Flame and patch perthite may be explained similarly, or may form by the replacement of one feldspar by another.

Zoning. Zoning is present in some alkali-feldspars but is most common in plagioclase, to which the following descriptions apply.

Normal zoning is a progressive increase in Na content from the core to the rim of a plagioclase crystal. It is very common in igneous rocks, and is explained as a natural consequence of crystallisation as defined by the solidus and liquidus curves of Fig. 103 (see Turner and Verhoogen, 1960, or other petrology texts for a full explanation).

Reverse zoning is a progressive increase in Ca content from the core to the rim of a plagioclase, and is less common than normal zoning. In metamorphic rocks it may be explained by crystallisation during a progressive increase in metamorphic grade.

Oscillatory zoning (Fig. 121) consists of a sequence of normal zones which are separated by sharp reversals in composition. There is usually an overall normal trend in the zoning. The most satisfactory explanation is that given by Bottinga et al. (1966), who ascribe its development to a combination of crystal growth mechanisms and the relative rates of diffusion of Al and Si from the magma to the crystal faces. Well-developed oscillatory zoning is characteristic, if not diagnostic, of magmatic crystallisation.

Fig. 121. Oscillatory zoning in broken plagioclase crystals, Te Pua Andesite, New Zealand (view measures 0.5 mm × 0.5 mm). Crossed-polarised light.

Myrmekite. This is a microscopic intergrowth of plagioclase and rods of quartz. The quartz rods are circular in cross-section and are often arranged in a radiating and branching manner perpendicular to the outer boundary of

 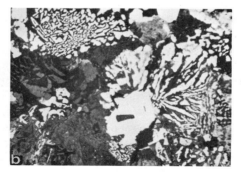

Fig. 122. (a) Myrmekite, Constant Gneiss, New Zealand (view measures 0.8 mm × 0.6 mm). (b) Granophyric intergrowth, Pepin Island, New Zealand (view measures 2.4 mm × 1.7 mm). Crossed-polarised light.

the plagioclase (Fig. 122a). The quartz in any one myrmekite has a single optical orientation. Myrmekite is almost ubiquitous in granites and granite-gneisses, and normally occurs along the grain boundaries of alkali-feldspar which it appears to replace. Interpretations of myrmekite include: replacement of K in alkali-feldspar by Na and Ca which causes a release of SiO_2 (Becke's hypothesis — see Ashworth, 1972); exsolution of Schwantke's molecule $Ca(AlSi_3O_8)_2$ and $NaAlSi_3O_8$ to the grain boundaries of the alkali-feldspar, the SiO_2 being released from Schwantke's molecule (Schwantke's hypothesis — see Ashworth, 1972); porphyroblastic growth of plagioclase including a recrystallising ground mass of quartz (Shelley, 1973).

Granophyric intergrowth. An intergrowth of quartz and K—Na-feldspar in which the quartz rods (which have a single optical orientation over large areas) are coarser than in myrmekite. The rods mostly have triangular cross-sections and may radiate within the feldspar crystal (Fig. 122b). It is generally interpreted (Barker, 1970) as resulting from the simultaneous crystallisation of quartz and alkali-feldspar from magma at low pressures (hypersolvus conditions — see Fig. 102a). Some superficial granite intrusions (grano-phyres) are composed mainly of this intergrowth.

Graphic intergrowth. This is similar to granophyric intergrowth. However, the quartz rods are coarser, clearly visible in hand specimen, closely parallel to each other, and have cuneiform cross-sections. The feldspar may be a K-rich alkali-feldspar or a sodic plagioclase. The intergrowth is generally interpreted (Barker, 1970) as the result of the simultaneous crystallisation of quartz and feldspar at higher pressures (subsolvus conditions — see Fig. 102b). It occurs in some granites, but is most common in pegmatites.

The feldspathoids

As the name implies, the feldspathoids are similar in composition to the feldspars. They are related to them by the equation:

Feldspathoid + silica = feldspar

Feldspathoids and the silica minerals are antipathetic, although feldspathoids commonly coexist with the feldspars.

No. 75. LEUCITE (+ve, pseudocubic) $K[AlSi_2O_6]$

Tetragonal (pseudocubic), cubic above 625° C. Very poor {110} cleavages. Equant crystals; euhedral—subhedral trapezohedral crystals with eight-sided sections. Complex transformation-twin patterns on {110} very common (Fig. 123).

Colour in thin section: colourless.

Fig. 123. Leucite with weak birefringence and complex twin patterns, leucitophyre, Eifel, Germany (view measures 2.6 mm × 1.7 mm). Crossed-polarised light.

Optical properties: uniaxial +ve or isotropic.
ω = 1.508—1.511, ϵ = 1.509—1.511.
δ = 0.00—0.001.
Small crystals may appear isotropic, but large crystals nearly always display a very weak δ and complex twin patterns.

Occurrence. As phenocrysts, or less commonly in the groundmass of silica-poor K-rich volcanic rocks, often with other feldspathoids such as sodalite. May be replaced by a mixture of alkali-feldspar and nepheline (called pseudoleucite).

Distinguishing features. The low relief (R.I. < Canada balsam), very low δ, twin patterns, crystal shape (eight-sided sections), and occurrence are distinctive. It does not occur with quartz. Leucite may be confused with sodalite and analcite, but both these minerals lack the distinctive twinning; in addition, the R.I. of analcite is lower than that of leucite, and sodalite has a different crystal shape.

No. 76. NEPHELINE (—ve) $(K,Na)Na_3[Al_4Si_4O_{16}]$

Kaliophilite, the K-rich analogue of nepheline is rare. Somewhat less rare is the polymorph *kalsilite* ($KAlSiO_4$).

Hexagonal. Poor $\{10\bar{1}0\}$ cleavages. Euhedral—subhedral crystals, usually very stubby prisms with a hexagonal cross-section and rectangular (almost square) side-sections; also anhedral.

Colour in thin section: colourless.

Optical properties: uniaxial —ve.
ω = 1.529—1.547, ϵ = 1.526—1.542.
δ = 0.003—0.007.
Square or rectangular sections of nepheline have straight extinction, and hexagonal sections provide uniaxial-cross interference figures; cleavage rarely seen.

Occurrence. In a wide range of silica-poor plutonic and volcanic igneous rocks, e.g. nepheline-syenite and phonolite. Also in nepheline-syenites and associated nepheline-bearing gneisses that are generally believed to have been produced by the metasomatic activity known as "fenitisation". Nepheline is also formed in reaction zones between carbonates and basic igneous rocks. Kalsilite is present in some K-rich volcanic rocks. Nepheline alters easily to "sericite", zeolites, sodalite and cancrinite.

Distinguishing features. The R.I. close to canada balsam, the low or very low δ, the uniaxial —ve character, lack of good cleavage, and alteration are distinctive. It does not occur with quartz. Often associated with alkali-feldspar with which it may be confused, but feldspar is biaxial and has good cleavages. Na-scapolite is similar, but has a higher δ and better cleavages. In fine-grained rocks, the presence of nepheline, or the *amount* of nepheline present, may be difficult to ascertain, and a stain test using phosphoric acid and methylene-blue dye may be used (Shand, 1939). Kalsilite and nepheline are best distinguished using X-rays.

No. 77. CANCRINITE (—ve) $(Na,Ca,K)_{6-8}[Al_6Si_6O_{24}](CO_3,SO_4,Cl)_{1-2} \cdot 1-5H_2O$

Hexagonal. Perfect $\{1010\}$ cleavages. Usually anhedral.

Colour in thin section: colourless.

Optical properties: uniaxial —ve.
ω = 1.490—1.528, ϵ = 1.488—1.503.
δ = 0.002—0.025.
R.I. and δ decrease with increasing SO_4.
Crystals have straight extinction and are fast parallel to good cleavage traces.

Occurrence. Forms as a late-stage magmatic mineral in silica-poor plutonic igneous rocks such as nepheline-syenite. Also as a secondary mineral replacing nepheline and feldspar in such rocks.

Distinguishing features. The low R.I., uniaxial —ve character, good prismatic cleavages and occurrence are distinctive. Moderately birefringent cancrinite may superficially resemble muscovite, but cancrinite differs in its lower R.I. and length-fast nature. Scapolite has a higher R.I. and less distinct cleavage.

No. 78. SODALITE GROUP (isotropic)

Sodalite	$Na_8[Al_6Si_6O_{24}]Cl_2$
Nosean	$Na_8[Al_6Si_6O_{24}]SO_4$
Haüyne	$(Na,Ca)_{4-8}[Al_6Si_6O_{24}](SO_4,S)_{1-2}$

Cubic. Poor $\{110\}$ cleavages. Euhedral—subhedral dodecahedral crystals with six-sided sections; also anhedral.

Colour in thin section: colourless, pale pink, blue or gray; sometimes colour-zoned.

Optical properties: isotropic.
$n = 1.483–1.487$ (sodalite), $1.461–1.495$ (nosean), $1.493–1.509$ (haüyne).
Crystals are often characterised by a zonal pattern of inclusions, some zones appearing dark and cloudy in thin section.

Occurrence. Members of the sodalite group are found in silica-poor igneous rocks such as nepheline-syenites and phonolites, often associated with nepheline and leucite. They also occur in metamorphosed and metasomatised carbonate rocks near igneous contacts.

Distinguishing features. The low R.I., isotropic character, crystal shape (six-sided sections) and occurrence are distinctive. May be confused with analcite and leucite, but both these minerals, if euhedral, have eight-sided sections; in addition, leucite has a higher R.I., a weak δ, and twinning. Anhedral analcite may be difficult to distinguish, although analcite has a different and better cleavage. Microchemical tests can be used to distinguish the members of the sodalite group, and to distinguish these from analcite (Deer et al., 1963, vol. 4, pp. 297 and 345).

The zeolites

The zeolites are hydrated silicates of aluminium and the alkalies or alkaline-earths. When heated, they readily loose some or all of the water without destruction of the structural framework; this water can be readily reabsorbed. There are numerous zeolites, and only a few are described here.

No. 79. NATROLITE (+ve) $Na_2[Al_2Si_3O_{10}] \cdot 2H_2O$

No. 80. MESOLITE (+ve) $Na_2Ca_2[Al_2Si_3O_{10}]_3 \cdot 8H_2O$

No. 81. SCOLECITE (−ve) $Ca[Al_2Si_3O_{10}] \cdot 3H_2O$

No. 82. THOMSONITE (+ve) $NaCa_2[(Al,Si)_5O_{10}]_2 \cdot 6H_2O$

No. 83. HEULANDITE (+ve) $(Ca,Na_2)[Al_2Si_7O_{18}] \cdot 6H_2O$

   (including *clinoptilolite*)

No. 84. STILBITE (−ve) $(Ca,Na_2,K_2)[Al_2Si_7O_{18}] \cdot 7H_2O$

No. 85. ANALCITE (isotropic) $Na[AlSi_2O_6] \cdot H_2O$

No. 86. WAIRAKITE (+ve or −ve) $Ca[Al_2Si_4O_{12}] \cdot 2H_2O$

No. 87. CHABAZITE GROUP (−ve or +ve) $Ca[Al_2Si_4O_{12}] \cdot 6H_2O$

   (including *gmelinite* and *levyne*)

TABLE VIII

Properties of the zeolites

Mineral	Cyrstal system and cleavages	Minimum R.I.	Maximum R.I.	δ	2V, sign and orientation
Natrolite	Orthorhombic. Two distinct {110} cleavages	1.473 (α)	. 1.496 (γ)	ca. 0.012	+ve, 2V = 58°–64°. X = a, Y = b, Z = c
Mesolite	Monoclinic, β = ca. 90°. Two perfect {101} cleavages	1.504 (β)	1.508 (β)	0.001	+ve, 2V = 80°. X∧c = 8°, Y = b
Scolecite	Monoclinic, β = ca. 90°. Two distinct {110} cleavages	1.507 (α)	1.521 (γ)	0.007–0.010	–ve, 2V = 36°–56°. X∧c = 18°, Z = b
Thomsonite	Orthorhombic. Perfect {010} and distinct {100} cleavages	1.497 (α)	1.544 (γ)	0.006–0.016	+ve, 2V = 38°–75°. X = a, Y = c, Z = **b**
Heulandite	Monoclinic, β = 91°. Perfect {010} cleavage	1.476 (α)	1.512 (γ)	0.002–0.008	+ve, 2V = 0°–74°. X∧a = 0°–34°, Z = b
Stilbite	Monoclinic, β = 129°. Distinct {010} cleavage	1.482 (α)	1.513 (γ)	0.008–0.014	–ve, 2V = 28°–49°. X∧c = ca. 5°, Y = b
Analcite	Cubic. Poor cubic cleavage	1.479 (n)	1.493 (n)	0.00–0.001	
Wairakite	Monoclinic, pseudocubic. Intersecting twins on {110}	1.498 (α)	1.502 (γ)	0.004	+ve or –ve, $2V_z$ = 70°–105°. X = ca. b, Y = ca. a, Z = ca. c.
Chabazite, gmelinite, and levyne	Trigonal. Poor {10$\bar{1}$1} cleavages (pseudocubic cleavage)	1.470	1.505	0.002–0.015 usually <0.005	–ve or +ve (often anomalously biaxial), 2V = 0°–32°
Laumontite	Monoclinic, β = 111°. Three distinct cleavages, {110} and {010}	1.502 (α)	1.526 (γ)	0.010–0.015	–ve, 2V = 26°–47°. Y = b, Z∧c = 8°–33°
Phillipsite	Monoclinic, β = ca. 90°. Distinct {010} and {100} cleavages. Common interpenetrant twinning	1.483 (α)	1.514 (γ)	0.003–0.010	+ve, 2V = 60°–80°. X = b, Z∧c = 10°–29°

Note: All zeolites are colourless in thin section. The properties of the zeolites may change if heated (during thin-section making, etc.).

No. 88. LAUMONTITE (—ve) $Ca[Al_2Si_4O_{12}] \cdot 4H_2O$

No. 89. PHILLIPSITE (+ve) $(\frac{1}{2}Ca,K,Na)_3[Al_3Si_5O_{16}] \cdot 6H_2O$

Crystallography and optical properties: the principal properties of the zeolites are compiled in Table VIII.

Orientation diagrams and habit. Natrolite, mesolite, scolecite and thomsonite are often fibrous. Natrolite (Fig. 124a), mesolite (Fig. 124b), and thomsonite (Fig. 124d) have straight extinction; natrolite is always length slow whereas mesolite and thomsonite have the O.A.P. across the length, and may be length fast or slow. Scolecite (Fig. 124c) has inclined extinction with $X \wedge c = 18°$.

Heulandite and stilbite are characterised by a single cleavage and platy habit with tabular or sheaf-like crystals parallel to (010). Heulandite (Fig. 124e) has the O.A.P. across the cleavage whereas stilbite (Fig. 124f) has the O.A.P. parallel to (010).

Analcite is cubic, and euhedral crystals are eight-sided in thin section. It is often anhedral. Some analcite may display a weak birefringence. Wairakite, essentially a Ca-analcite, is pseudocubic, and usually has interpenetrating twin lamellae.

Members of the chabazite group are normally uniaxial —ve, but they may be +ve or anomalously biaxial. They crystallise most commonly with prominent $\{10\bar{1}1\}$ forms which resemble cubes. The members chabazite, gmelinite and levyne are best sorted with X-rays.

Laumontite (Fig. 124g) is characterised by three good cleavages seen together in (001) sections, and a small extinction angle $Z \wedge c$.

Phillipsite (Fig. 124h) commonly has interpenetrant twinning. It has inclined extinction with $Z \wedge c = 10° - 29°$.

Occurrence. There are three principal modes of occurrence:

(1) *Primary igneous.* Analcite is the only zeolite to form as a primary-igneous mineral (e.g. in some basalts, teschenites, and related rocks). It forms late, and is difficult to distinguish from secondary analcite.

(2) *Secondary in igneous rocks.* Zeolites most typically occur as secondary minerals in amygdales and fissures, chiefly in basic volcanic rocks. Zonal sequences are known, so that in the Tertiary lavas of East Iceland, there is a zeolite-free zone at the top with successive zones downwards of chabazite—thomsonite, analcite, and mesolite—scolecite (Walker, 1960). Other zonal sequences have been described. The zeolites may replace feldspars and nepheline in igneous rocks.

(3) *Diagenetic, in low-grade metamorphic rocks and hot-spring deposits.* Many of the zeolites form during diagenesis in both marine (including modern deep-ocean sediments) and non-marine sediments; these zeolites commonly replace glassy tuffaceous material. In addition, zeolites may replace tuffaceous material, detrital plagioclase, and fossil material to form zonal sequences in response to burial metamorphism (the "zeolite facies") or hot-spring activity (Coombs, 1971). In zonal sequences, analcite and heulandite are succeeded with increasing burial by laumontite, and with increasing temperature (in hot-spring deposits) by wairakite.

Distinguishing features. The zeolites are quite distinctive as a group. All are colourless, they have a low relief with their R.I. commonly being less than that of canada balsam, and their δ is low or very low. Their typical occurrence as secondary minerals in amygdales and fissures is distinctive.

Analcite may be confused with leucite or sodalite, but leucite has a higher R.I. and

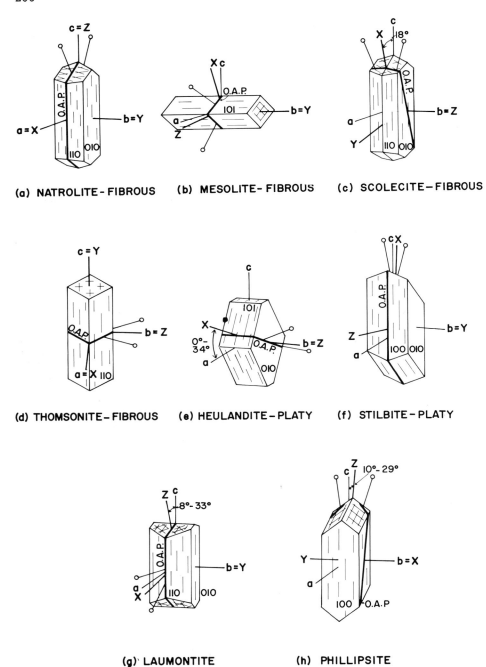

Fig. 124. Optical orientation of the zeolites.

distinctive twinning, and euhedral sodalite crystals are six-sided. Microchemical tests (see sodalite) may be used to distinguish these minerals.

Laumontite and scolecite have similar properties to the alkali-feldspars, but the cleavages of laumontite and the fibrous nature of scolecite are distinctive.

Thomsonite is similar in properties to gypsum, but the straight extinction of thomsonite is distinctive.

Refer to Table VIII and Fig. 124 for distinctions amongst the zeolites.

No. 90. SCAPOLITE (—ve)

A series between *marialite* $3Na[AlSi_3O_8] \cdot NaCl$ and *meionite* $3Ca[Al_2Si_2O_8] \cdot CaCO_3$. K may also be present. Pure end-members are not known naturally.

Tetragonal. Distinct $\{100\}$ and imperfect $\{110\}$ cleavages. Long prismatic crystals or anhedral, granular.

Colour in thin section: colourless.

Optical properties: uniaxial —ve.
$\omega = 1.546$—1.600, $\epsilon = 1.540$—1.571.
$\delta = 0.006$—0.036.
R.I. and δ increase with meionite (Ca) content (Fig. 125).
Crystals are fast parallel to cleavages and crystal outlines with straight extinction.

Occurrence. Forms instead of plagioclase in a wide range of regionally metamorphosed rocks under high CO_2 pressures, or in the presence of abundant brine (NaCl). Also forms metasomatically in metamorphic rocks where NaCl is introduced, and in altered igneous rocks affected by pneumatolytic activity. Commonly found in skarns at the contact of igneous intrusions and carbonate rocks.

Distinguishing features. Low birefringent scapolite superficially resembles a number of minerals. However, it is usually distinguished by its uniaxial —ve character and straight extinction. Cancrinite has a lower R.I. than scapolite, and a higher δ than low R.I. scapolite. Nepheline and beryl differ in their habit and lack of good cleavage.

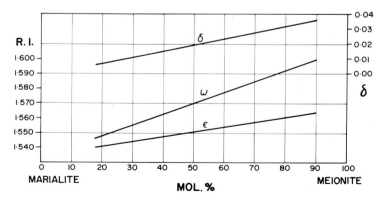

Fig. 125. Variation of R.I. and birefringence in scapolite. Based on data in Deer et al. (1963).

F. VOLCANIC GLASS

Volcanic glass, of course, is not a mineral. However, it is a substance commonly encountered in microscope work, and a description of it is quite relevant to the purpose of this book.

No. 91. VOLCANIC GLASS (isotropic)

Amorphous. May be massive, vesicular, or show perlitic cracking. Some volcanic rocks are made entirely of glass, but others contain glass as a groundmass to crystals of feldspar, pyroxene, etc. Many glasses are more or less devitrified to felsite, spherulites, and/or crystallites of various kinds.

Colour in thin section: usually colourless, grey, brown or red.

Optical properties: isotropic.
$n = 1.485-1.62$ (Fig. 126).
R.I. generally increases with a decrease in SiO_2 content of the glass (Fig. 126). However, many other factors (e.g. H_2O content) also control R.I., and a close correlation of SiO_2 with R.I. is not possible. Mathews (1951) has shown that good correlations between R.I. and SiO_2 can be made for glasses from particular rock suites produced by artificial fusion. Curves for three suites of artificially fused glass are given in Fig. 126.

Occurrence. In lavas and hypabyssal igneous intrusions such as dykes, occasionally making up the bulk of the rock as in obsidian, but more frequently forming the groundmass and containing abundant small crystals and phenocrysts. Also as shards, scoria and other fragments in pyroclastic rocks. In addition to devitrification products, glass may be altered to "palagonite", a green, waxy, hydrated substance, or replaced by zeolites.

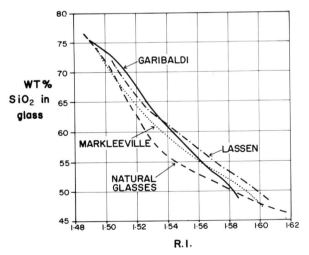

Fig. 126. R.I. variation with SiO_2 content of volcanic glass for natural examples and three suites of fused glass. Redrawn with permission from Mathews (1951, fig. 2).

Distinguishing features. The occurrence and isotropic character are distinctive. The R.I. is higher than that of opal.

G. NON-SILICATES

Of the non-silicates, only the carbonates, iron oxides, the bauxite minerals, and the various salts found in evaporites occur as the prime constituents of important rock masses. Many of the other non-silicates are important as accessory minerals in a wide range of rock types. They are described here under the headings: elements; oxides and hydroxides; sulphides; halides; sulphates; carbonates; and phosphates. Minerals such as galena, which are important economic minerals but unimportant in petrology, are omitted.

(a) Elements

No. 92. GRAPHITE (opaque)   C

Hexagonal. Perfect $\{0001\}$ cleavage. Often very fine grained; crystals are hexagonal plates elongate parallel to the cleavage. Very soft, with a soapy feel.

Colour in thin section: opaque; black with a metallic lustre in reflected light.
Tabular crystals with perfect cleavage are often crinkled in a manner similar to micas.

Occurrence. In slates, schists, gneisses, marbles, and veins associated with these rocks; often finely disseminated. The grey colour of marble may be due to the presence of fine-grained graphite. Also rarely found as an accessory mineral in plutonic and volcanic igneous rocks.

Distinguishing features. The opaque character, black colour, platy crystals, and softness are distinctive. May be confused with magnetite or ilmenite, but both these minerals are harder; in addition, magnetite is magnetic and lacks cleavage, and ilmenite plates are not commonly crinkled or bent. Ilmenite may be altered to white (opaque) leucoxene.

(b) Oxides and hydroxides

No. 93. PERICLASE (isotropic)   MgO

Cubic. Perfect cubic cleavage. Crystals may be euhedral cubes or octahedra; also as anhedral grains.

Colour in thin section: colourless.

Optical properties: isotropic.
$n = 1.736$—ca. 1.745.

Occurrence. In medium- or high-grade thermally metamorphosed marbles and dolomites, especially near igneous contacts where dolomite has dissociated. Commonly present only as relic patches in grains altered to brucite.

Distinguishing features. The high relief, isotropic character, perfect cubic cleavage, occurrence, and alteration to brucite are distinctive.

No. 94. CORUNDUM (−ve) Al_2O_3

Ruby (red) and *sapphire* (blue) are precious varieties.

Trigonal. No cleavage, but well-developed $\{0001\}$ parting common. Euhedral crystals are often tapering pyramids ("barrel-shaped") with hexagonal cross-sections and the $\{0001\}$ form developed; also anhedral, granular. Simple or lamellar twinning with $\{10\bar{1}1\}$ composition planes common.

Colour in thin section: colourless, or pale blue, green, yellow, or pink; often colour-zoned or with an irregular distribution of colours; weak to strong pleochroism with $\omega > \epsilon$.

Colour in detrital grains: as above but deeper colours, and often with marked pleochroism.

Optical properties: uniaxial −ve; may be anomalously biaxial.
$\omega = 1.765-1.772$, $\epsilon = 1.759-1.763$.
$\delta = 0.005-0.009$.
Fe-rich corundum with $\omega = 1.794$ and $\epsilon = 1.785$ has been reported.

Occurrence. Found in nepheline-syenites and syenites, and associated pegmatites; also in quartz-free plagioclase-rich "dykes" cutting mafic or ultramafic rocks. Al-rich xenoliths in igneous rocks may contain corundum, often associated with spinel. Corundum also occurs relatively rarely in a wide variety of metamorphic rocks, especially Al-rich pelites, some carbonates and rocks desilicated near igneous intrusions. It is the prime constituent of emery (metamorphosed bauxite). Locally, it may be important as a detrital mineral. Corundum alters most commonly to micas, diaspore, gibbsite or other Al-rich minerals.

Distinguishing features. The high relief combined with a low δ and uniaxial −ve character is distinctive. Corundum is also distinguished by its hardness in hand specimen.

No. 95. CASSITERITE (+ve) SnO_2

Tetragonal. Poor $\{100\}$ and $\{110\}$ cleavages. Euhedral crystals may be short prisms with pyramidal terminations, elongate prisms or acicular; also subhedral or anhedral. Twins common on $\{011\}$ (geniculate or "knee-like" twinning).

Colour in thin section: colourless, yellow, brown or red; pleochroism weak or strong with $\epsilon > \omega$.

Optical properties: uniaxial +ve.
$\omega = 1.990-2.010$, $\epsilon = 2.093-2.100$.
$\delta = 0.090-0.103$.
Crystals have extinction parallel to prism faces but oblique to twin planes; the length-slow character is difficult to determine because of the very high δ.

Occurrence. Most commonly found together with W-, Li-, B- and F-rich minerals in granite pegmatites, and associated greisen and veins; also as an accessory mineral in some granites. Locally it may be important as a detrital mineral.

Distinguishing features: The extreme relief, very high o, and uniaxial +ve character are distinctive. May be confused with rutile and occasionally zircon, but the δ of rutile is higher and that of zircon lower; these distinctions may be difficult to make in thick detrital grains which display indeterminable very high-order interference colours.

No. 96. RUTILE (+ve) TiO_2

Tetragonal. Distinct $\{110\}$ and $\{100\}$ cleavages. Euhedral crystals are short to elongate or acicular prisms; less commonly anhedral, granular. Twins common on $\{011\}$ (geniculate or "knee-like" twinning).

Colour in thin section: usually red, red-brown, or yellowish; often weak but sometimes strong pleochroism with $\epsilon > \omega$.

Colour in detrital grains: usually deep colours as above, often almost opaque.

Optical properties: uniaxial +ve; sometimes anomalously biaxial.
ω = 2.605—2.616, ϵ = 2.899—2.903.
δ = 0.286—0.294.
Crystals are characterised by indeterminable very high-order interference colours which may be confused with 1st-order white. Insertion of the sensitive-tint plate will not effect any observable change in the very high-order colours. Prismatic crystals have straight extinction.

Occurrence. A widely distributed accessory mineral, especially in metamorphic rocks and less commonly in igneous rocks. Usually as small prismatic crystals. Rutile needles are commonly embedded in other minerals forming regular patterns parallel to crystal faces, especially in amphibole and chlorite resulting from the alteration of pyroxene. Also found in veins, and as a common detrital mineral. May be altered to white (opaque) leucoxene.

Distinguishing features. The extreme relief, very high δ, crystal habit, and colour are distinctive. May be confused with the less common minerals cassiterite, brookite and anatase; the δ of these three minerals is lower although this may be difficult to determine in thick detrital grains. Anatase is —ve, and brookite has indistinct extinction due to its high dispersion.

No. 97. ANATASE (—ve) TiO_2

Tetragonal. Perfect $\{001\}$ and $\{111\}$ cleavages. Pyramidal crystals resembling octahedra.

Colour in thin section: yellow, brown, blue or black; usually weak pleochroism, $\omega < \epsilon$ or $\omega > \epsilon$.

Colour in detrital grains: deep colours as above.

Optical properties: uniaxial —ve.
ω = 2.561, ϵ = 2.488.
δ = 0.073.

Occurrence. Principally as a detrital or diagenetic mineral, sometimes prominent in heavy-mineral suites. Also in veins and altered igneous and metamorphic rocks. May be altered to white (opaque) leucoxene.

Distinguishing features. The extreme relief, colour, very high δ, shape and cleavages are characteristic. Distinguished from rutile and brookite by its —ve character and lower δ. In addition, brookite differs in its indistinct extinction.

No. 98. BROOKITE (+ve) TiO_2

Orthorhombic. No good cleavage. Crystals are often tabular parallel to $\{100\}$ and slightly elongate parallel to c.

Colour in thin section: pale to dark yellow or brown; pleochroism absent or weak.

Colour in detrital grains: deeper colours as above.

Optical properties: biaxial +ve. $2V_z = 0° - 30°$.
$\alpha = 2.583$, $\beta = 2.584 - 2.586$, $\gamma = 2.700 - 2.741$.
$\delta = 0.117 - 0.158$. $X = c$ or a, $Y = a$ or c, $Z = b$.
Brookite has a very strong dispersion which causes incomplete extinction.

Occurrence. Principally as a detrital or diagenetic mineral, sometimes prominent in heavy-mineral suites. Also in veins and altered igneous and metamorphic rocks. May be altered to white (opaque) leucoxene.

Distinguishing features. The extreme relief, colour, very high δ and incomplete extinction are distinctive. May be confused with rutile and anatase, but both these minerals usually extinguish completely, rutile has a higher δ, and anatase is —ve.

No. 99. PEROVSKITE (+ve, pseudo-isotropic) $CaTiO_3$

There may be substitution of Nb for Ti, and Na, Fe^{2+}, or the rare-earths for Ca.

Monoclinic? (pseudocubic). Pseudocubic cleavage, not usually observed in small crystals. Crystals may be cubic or octahedral in habit. Complex multiple twinning on $\{111\}$ common.

Colour in thin section: pale to very dark brown, rarely grey or green (Nb-rich varieties); pleochroism absent or very weak with $Z > X$.

Optical properties: isotropic, or biaxial with $2V = $ ca. $90°$.
$n = 2.30 - 2.38$.
$\delta = 0.00 - 0.002$.

Occurrence. An accessory mineral in some feldspathoidal or melilite-bearing igneous rocks, and in some carbonatites. Also found in thermally metamorphosed carbonate rocks.

Distinguishing features. The extreme relief and very low or zero δ are distinctive. Garnet has a lower relief and a different crystal habit. Very dark, almost opaque varieties of perovskite may be confused with the other opaque minerals.

No. 100. SPINEL GROUP (isotropic or opaque)

The spinel group can be divided into the following three series:

No. 100A. Spinel series $(Mg,Fe^{2+},Zn,Mn)Al_2O_4$
No. 100B. Chromite series $(Fe^{2+},Mg)Cr_2O_4$
No. 100C. Magnetite series $(Fe^{2+},Mg,Zn,Mn,Ni)Fe_2^{3+}O_4$

Cubic. No cleavage. Euhedral crystals are octahedra; also anhedral. Twinning with $\{111\}$ twin planes — not usually visible in thin section.

Spinel series

Includes *spinel* $MgAl_2O_4$, *pleonaste* $(Mg,Fe^{2+})Al_2O_4$, *hercynite* $Fe^{2+}Al_2O_4$ and *gahnite* $ZnAl_2O_4$; *picotite* is Cr-rich hercynite. Pure end-members are rare.

Colour in thin section: colourless or a wide variety of colours, but especially blue and brown (spinel), green or blue-green (pleonaste), dark green to almost black and opaque (hercynite) and dark blue-green (gahnite).

Optical properties: isotropic.
n varies in the range ca. 1.715 (spinel) — 1.80 (gahnite) — 1.835 (hercynite) — 1.98 (picotite).
Diamond- or square-shaped sections common.

Occurrence. Spinel and pleonaste are most commonly found in high-grade metamorphosed carbonate rocks often associated with chondrodite, phlogopite or forsterite, and in Al-rich, Si-poor schists and xenoliths. Hercynite is found in more Fe-rich aluminous schists, and may occur with quartz in Si-rich granulites. Hercynite and pleonaste are found rarely as accessories in basic and ultramafic igneous rocks. Picotite occurs in ultramafic rocks, and gahnite in granite pegmatites and Zn ores. Members of the spinel series are found intergrown with hornblende in corona structures between olivine and plagioclase in gabbroic rocks.

Distinguishing features. The habit, high relief, isotropic nature and colours are distinctive. Periclase differs in its perfect cleavage, and garnet has a different habit and is more commonly pink than spinel.

Chromite series

Chromite (sensu stricto) is $Fe^{2+}Cr_2O_4$, but most natural crystals contain substantial Mg replacing Fe^{2+}.

Colour in thin section: usually dark brown or opaque; opaque grains are usually translucent at their edges.

Optical properties: isotropic.
n = ca. 2.00—2.16.
Diamond- or square-shaped sections common.

Occurrence. In ultramafic rocks such as dunite, peridotite, and serpentinite, sometimes concentrated in layers. Locally it may be important in detrital heavy-mineral suites.

Distinguishing features. The deeply coloured almost opaque nature, crystal habit and isotropism are distinctive. Deeply coloured melanite-garnet has a different habit and occurrence.

Magnetite series

Magnetite ($Fe^{2+}Fe_2^{3+}O_4$) and varieties with some replacement of Fe^{2+} by Mg are the common members of this series. Some Ti may be present.

Colour in thin section: opaque; black with a metallic lustre in reflected light.
Diamond- or square-shaped sections; also commonly granular.

Occurrence. Very abundant as an accessory mineral or as a principal constituent in a wide variety of igneous and metamorphic rocks, especially basic igneous rocks. Also a common detrital mineral, sometimes concentrated by stream or tidal action to form magnetite sands.

Distinguishing features. The opacity, black colour and metallic lustre in reflected light, strongly magnetic character, and crystal habit (if developed) are distinctive. May be confused with graphite or ilmenite, but neither of these minerals is strongly magnetic, and both commonly crystallise as hexagonal plates. In addition, graphite is very soft, and ilmenite may be altered to white (opaque) leucoxene.

No. 101. ILMENITE (opaque) $FeTiO_3$

Trigonal. No good cleavage. Euhedral crystals are thin hexagonal plates or tablets; often skeletal crystals; also anhedral, granular.

Colour in thin section: opaque; black with a metallic lustre in reflected light.
Euhedral crystals appear as elongate plates in thin section. Ilmenite is commonly altered to opaque *leucoxene* (TiO_2) which is white in reflected light (often looks like cotton-wool).

Occurrence. A common accessory mineral in a wide range of igneous and metamorphic rocks, especially mafic and ultramafic types. Often found with magnetite. A common detrital mineral, sometimes concentrated by stream or tidal action to form ilmenite sands.

Distinguishing features. The opacity, black colour and metallic lustre in reflected light and crystal habit (if developed) are distinctive. May be confused with graphite and magnetite, but graphite is very soft and often crinkled, and magnetite has a different habit and is strongly magnetic. The alteration of ilmenite to leucoxene is distinctive.

No. 102. HEMATITE (−ve, often opaque) Fe_2O_3

Trigonal. No good cleavage. Flaky with $\{0001\}$ well-developed or rhombohedral crystals; commonly anhedral, massive, earthy.

Colour in thin section: occasionally red, translucent, especially at the edges of grains; commonly opaque and black with a metallic lustre in reflected light.

Optical properties: opaque or uniaxial −ve.
$\omega = 3.15-3.22$, $\epsilon = 2.87-2.94$.
$\delta = 0.28$.

Occurrence. Rare as a primary mineral in igneous rocks, but common as an alteration product and fumarole deposit. Very common in some metamorphosed Fe-rich rocks, especially the Precambrian banded iron ores in which it occurs as bands alternating with quartz. Hematite is a common red colouration and cement in sediments, and may be present in soils. Also found in veins.

Distinguishing features. The extreme relief, very high δ and red colour of translucent hematite are distinctive. Opaque hematite may be difficult to distinguish from magnetite, but hematite has a distinctive red streak (the streak of magnetite is black); magnetite is also strongly magnetic. Goethite, lepidocrocite and limonite have yellow or brown streaks, goethite is often fibrous, and limonite is isotropic.

No. 103A. GOETHITE (—ve)   FeO · OH

No. 103B. LEPIDOCROCITE (—ve)   FeO · OH

Limonite is amorphous or cryptocrystalline goethite or lepidocrocite with absorbed water.

Orthorhombic. Perfect {010} cleavage. Crystals of goethite are often fibrous parallel to c, whereas lepidocrocite is often platy parallel to {010} ; commonly massive, earthy, stalactitic, etc.

Colour in thin section: yellow or brown; the pleochroism of goethite is variable; lepidocrocite is strongly pleochroic with $Z > Y > X$.

Optical properties. *Goethite*: biaxial —ve. $2V_x = 0°—27°$.
$\alpha = 2.217—2.275$, $\beta = 2.346—2.409$, $\gamma = 2.356—2.415$.
$\delta = 0.139—0.140$. $X = b$, $Y = c$ or a, $Z = a$ or c.
Extreme dispersion so that the positions of Y and Z and the value of $2V$ change for different wavelengths of light.
Lepidocrocite: biaxial —ve. $2V_x = 83°$.
$\alpha = 1.94$, $\beta = 2.20$, $\gamma = 2.51$.
$\delta = 0.57$. $X = b$, $Y = c$, $Z = a$.
Limonite is isotropic with $n =$ ca. 2.0.

Occurrence. Both goethite and lepidocrocite form as common alteration products of Fe-bearing minerals under oxidising conditions. Found in soils, bogs, laterites, and sedimentary iron ores. Yellow ochre contains goethite, brown ochre contains lepidocrocite.

Distinguishing features. The extreme relief, yellow-brown colour and occurrence are distinctive. Goethite and lepidocrocite may be difficult to distinguish, although their birefringences differ. Limonite is distinguished by its isotropic character. Hematite may be most easily distinguished from all three by its red streak.

No. 104. BRUCITE (+ve)   Mg(OH)$_2$

Fe and Mn may substitute for some of the Mg.

Trigonal. Perfect {0001} cleavage. Crystals are very soft mica-like {0001} plates, or fibres elongate parallel to ω.

Colour in thin section: colourless.

Optical properties: uniaxial +ve; fibres may be anomalously biaxial.

ω = 1.559—1.59, ϵ = 1.579—1.60.
δ = 0.010—0.021.
Anomalous red-brown interference colours may replace 1st-order yellow-orange.
Plates and fibres are always length fast.

Occurrence. Most commonly found in thermally metamorphosed carbonate rocks, often as an alteration product of periclase. Also in veins cutting serpentinite and chlorite-rich rocks.

Distinguishing features. The soft, platy or fibrous crystals, low to moderate relief and δ, +ve character, and occurrence are distinctive. May resemble some micas, talc, chlorite, or serpentine, but the length-fast character of brucite differs from all these except +ve chlorite which is usually green and has a lower δ.

No. 105. GIBBSITE (+ve) $Al(OH)_3$

Also known as *hydrargillite*.

Monoclinic, β = 94°. Perfect $\{001\}$ cleavage. Hexagonal plates parallel to $\{001\}$; most commonly subhedral lamellae and in concretionary aggregates; usually very fine grained. Multiple twinning with $\{001\}$ composition planes common.

Colour in thin section: colourless or pale brown.

Optical properties: biaxial +ve. $2V_z$ usually <20°, changes on heating.
$\alpha = \beta$ = 1.565—1.571, γ = 1.580—1.595.
δ = 0.014—0.030. At room temperatures $X = b$, $Y \wedge a$ = 25°, $Z \wedge c$ = 21°.

Orientation diagrams. Crystals are usually too small for an exact determination of optical properties.
(001) section (Fig. 127a): off-centred acute bisectrix figure; δ' very low.
(010) section (Fig. 127b): obtuse bisectrix (flash) figure; δ' (moderate) usually ca. 0.02; inclined extinction with the angle between the fast direction and the good cleavage = 25°.
(100) section (Fig. 127c): off-centred flash figure; δ' (moderate) usually ca. 0.02; straight extinction and fast parallel to the cleavage.

Occurrence. One of the prime constituents of bauxite, usually with diaspore and boehmite, and in laterites and clays formed under similar conditions; also in emery deposits. Gibbsite occurs in veins and cavities in some Al-rich igneous rocks, and as an alteration product of corundum.

Distinguishing features. Superficially resembles brucite, the micas and clay minerals. The inclined extinction in (010) sections and the occurrence of gibbsite distinguish it from brucite, and the length-fast character distinguishes it from micas and the clay minerals. Gibbsite differs from associated diaspore and boehmite in its lower R.I.

No. 106. DIASPORE (+ve) $AlO(OH)$

Orthorhombic. Perfect $\{010\}$ cleavage, and other imperfect prismatic cleavages. Crystals are tabular parallel to $\{010\}$ and/or elongate parallel to c; often in massive fine-grained aggregates.

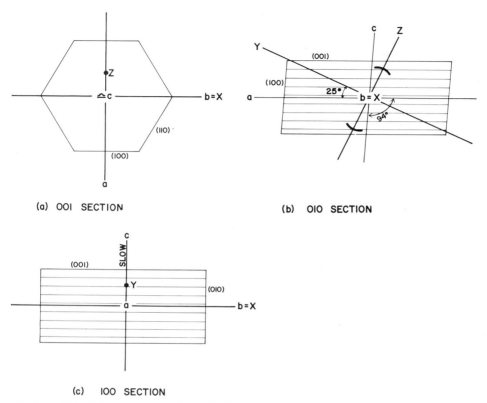

(a) OOI SECTION

(b) OIO SECTION

(c) IOO SECTION

Fig. 127. Orientation diagrams for gibbsite.

Colour in thin section: usually colourless, sometimes pink or brown pleochroic with $Z > Y > X$.

Optical properties: biaxial +ve. $2V_z = 84°-86°$.
$\alpha = 1.685-1.706$, $\beta = 1.705(?)-1.725$, $\gamma = 1.730-1.752$.
$\delta = 0.045-0.050$. $X = c$, $Y = b$, $Z = a$.
Crystals are usually length fast with the O.A.P. parallel to the length and good cleavage traces.

Occurrence. One of the prime constituents of bauxite, usually with gibbsite and boehmite, and in laterites and clays formed under similar conditions. Diaspore also occurs in emery deposits, in altered Al-rich rocks, and as an alteration product of Al-rich minerals such as corundum, kyanite and andalusite.

Distinguishing features. The high relief and δ, the O.A.P. parallel to the good cleavage, and occurrence are distinctive. May be confused with sillimanite or micas, but diaspore differs in its length-fast character, and it has a higher R.I. than the micas. Diaspore differs from associated gibbsite and boehmite in its higher R.I. and δ.

No. 107. BOEHMITE (+ve?) $AlO(OH)$

Orthorhombic. Distinct $\{010\}$ cleavage, rarely seen in crystals which are usually very fine-grained $\{001\}$ plates.

Colour in thin section: colourless.

Optical properties: biaxial +ve? $2V_z$ uncertain (moderate or large).
α = ca. 1.646—1.650, γ = ca. 1.661—1.662.
δ = ca. 0.012—0.015. Uncertain orientation.
Crystals have straight extinction and have been reported as length slow and fast.

Occurrence. One of the prime constituents of bauxite, usually with gibbsite and diaspore, and in laterites and clays formed under similar conditions.

Distinguishing features. The occurrence is distinctive. The fine-grained nature of boehmite has not allowed an exact characterisation of its optical properties. It differs from associated gibbsite in its straight extinction and higher R.I., and from diaspore in its lower R.I. and δ.

(c) Sulphides

No. 108A. PYRITE (opaque) FeS_2

No. 108B. MARCASITE (opaque) FeS_2

Pyrite is cubic, marcasite is orthorhombic. Neither has a good cleavage. Pyrite crystallises as cubes or pyritohedra (Fig. 5), marcasite as $\{010\}$ tablets, fibrous aggregates and concretions; pyrite may also be massive, granular.

Colour in thin section: opaque; brassy yellow with a metallic lustre in reflected light.

Occurrence. Pyrite has a wide range of occurrences, as a primary and secondary mineral in many igneous and metamorphic rocks, in skarns, in fumaroles, and as a diagenetic mineral in sediments, especially muds laid down in shallow water and reducing conditions. Pyrite is also common in veins and fissures. Marcasite is restricted to sediments and veins. Pyrite and marcasite readily alter to iron oxides, particularly limonite, sometimes accompanied by gypsum or other sulphates.

Distinguishing features. The opacity and brassy-yellow colour in reflected light are distinctive. Pyrite and marcasite may be confused with pyrrhotite and chalcopyrite. The euhedral cubes and pyritohedra of pyrite are distinctive, but massive material is less easy to identify. Pyrrhotite is magnetic and has a bronze colour in reflected light and chalcopyrite is softer and golden in colour. Marcasite is usually distinguished from pyrite by its generally fibrous nature and more restricted occurrence.

No. 109. PYRRHOTITE (opaque) $Fe_{1-x}S(x = 0—0.2)$

Monoclinic (pseudohexagonal), β = ca. $90°$. No good cleavage, but frequently has a $\{0001\}$ parting. Usually massive, granular or lamellar; euhedral crystals are hexagonal tablets.

Colour in thin section: opaque; bronze colour with a metallic lustre in reflected light.

Occurrence. As a primary mineral in some basic igneous rocks. More rarely found in thermally metamorphosed carbonate rocks, and some schists. Also forms as a diagenetic mineral in muds laid down under reducing conditions, and in veins. Pyrrhotite, like pyrite, alters to limonite.

Distinguishing features. Pyrrhotite resembles pyrite, but pyrite is brassy yellow in reflected light, and euhedral crystals differ in habit. Pyrrhotite, unlike pyrite, is magnetic, and will react to a hand-magnet. Chalcopyrite differs in its golden-yellow colour.

No. 110. CHALCOPYRITE (opaque) $CuFeS_2$

Tetragonal. No good cleavage. Usually massive; euhedral crystals are tetrahedral in shape.

Colour in thin section: opaque; golden-yellow with a metallic lustre in reflected light.

Occurrence. As a primary mineral in some basic igneous rocks, often associated with pyrite and pyrrhotite, and as a diagenetic or secondary mineral in sediments. Also in veins and a wide variety of metasomatic deposits.

Distinguishing features. The opacity and golden-yellow colour in reflected light are distinctive. It resembles pyrite and pyrrhotite, but pyrite is harder and has a brassy colour, and pyrrhotite is magnetic and has a bronze colour.

(d) Halides

No. 111. FLUORITE (isotropic) CaF_2

Cubic. Perfect {111} cleavages. Euhedral crystals are cubes; often anhedral, interstitial.

Colour in thin section: colourless or commonly blue or purple; often colour-zones or with an irregular distribution of colours.

Optical properties: isotropic.
$n = 1.433$—ca. 1.44, increases with rare-earth content.
The {111} cleavages are usually visible in thin section as two sets intersecting at ca. $70°$.
Crushed cleavage-bound fragments are triangular in outline.

Occurrence. Found as an interstitial accessory mineral (possibly of hydrothermal origin) in igneous rocks, especially acid and alkaline types; also in pegmatites and fumarole deposits. Fluorite is common in some hydrothermal veins, and also occurs as a detrital mineral and more rarely as a cement in sediments.

Distinguishing features. The low R.I., isotropism, perfect cleavages and purple colour (if present) are distinctive. Opal lacks cleavage and colour.

No. 112. HALITE (isotropic)   NaCl

Common rock-salt.

Cubic. Perfect cubic cleavage. Euhedral crystals are cubes; commonly massive, anhedral, granular. Soluble in water and has a salty taste.

Colour in thin section: colourless.

Optical properties: isotropic.
$n = 1.544$.

Occurrence. An evaporite from seawater or salt-lakes. May be found as scattered euhedra in some sediments, but most commonly occurs in massive beds, often intruded as massive salt-domes or walls.

Distinguishing features. The isotropism, very low relief, occurrence and solubility in water are distinctive. For the properties of associated salts see Table IX.

Nos. 113–118. OTHER WATER-SOLUBLE SALTS FOUND IN EVAPORITES

Halite is by far the most abundant of the water-soluble salts found in evaporites. The properties of halite and the other water-soluble salts most commonly encountered are summarised in Table IX, the data coming mainly from Borchert and Muir (1964) to which reference should be made for details of other evaporite minerals. Because of their solubility in water, they require special methods of thin-section preparation, and a suitable procedure has been described by Bennett (1958).

(e) Sulphates

The properties of a number of water-soluble sulphates found in evaporites are given in Table IX.

No. 119A. BARYTE (+ve) $BaSO_4$

No. 119B. CELESTINE (+ve) $SrSO_4$

Orthorhombic. Perfect $\{001\}$, good $\{210\}$ and distinct $\{010\}$ cleavages. Euhedral crystals are $\{001\}$ tablets which may be elongate parallel to a or b; often anhedral, granular. Baryte may have lamellar twinning on $\{110\}$.

Colour in thin section: colourless.

Optical properties. *Baryte*: biaxial +ve. $2V_z = 37°$.
$\alpha = 1.636$, $\beta = 1.637$, $\gamma = 1.648$.
$\delta = 0.012$. $X = c$, $Y = b$, $Z = a$.
Celestine: biaxial +ve. $2V_z = 51°$.
$\alpha = 1.622$, $\beta = 1.624$, $\gamma = 1.631$.
$\delta = 0.009$. $X = c$, $Y = b$, $Z = a$.

Orientation diagrams. *(100) section* (Fig. 128a): acute bisectrix figure; δ' (very low) <0.002; two cleavages at 90° to each other and straight extinction; extinction may be indistinct due to proximity of optic axes; O.A.P. across the best cleavage.
(001) section (Fig. 128b): obtuse bisectrix figure; δ' (low)$=$ 0 011 (baryte) or 0.007

TABLE IX

Properties of water-soluble salts found in evaporites

Mineral	Crystal system and cleavage	Taste	R.I. and δ	$2V$ and orientation
No. 112 Halite NaCl	Cubic. Perfect cubic cleavage	salty	$n = 1.544$	isotropic
No. 113 Sylvite KCl	Cubic. Perfect cubic cleavage	salty, bitter	$n = 1.490$	isotropic
No. 114 Carnallite $KCl \cdot MgCl_2 \cdot 6H_2O$	Orthorhombic. No cleavage	bitter	$\alpha = 1.466, \beta = 1.475$ $\gamma = 1.494, \delta = 0.028$	$2V_z = 66°$ $X = c, Y = b, Z = a$
No. 115 Langbeinite $K_2SO_4 \cdot 2MgSO_4$	Cubic. No cleavage	tasteless	$n = 1.534$	isotropic
No. 116 Kainite $4KCl \cdot 4MgSO_4 \cdot 11H_2O$	Monoclinic. Distinct $\{100\}$ and $\{110\}$ cleavages	salty, slightly bitter	$\alpha = 1.494, \beta = 1.505$ $\gamma = 1.516, \delta = 0.022$	$2V_x = 85°$ $Y = b, Z \wedge c = 1\overset{\circ}{3}°$
No. 117 Kieserite $MgSO_4 \cdot H_2O$	Monoclinic. Distinct $\{111\}$ and $\{110\}$cleavages	tasteless	$\alpha = 1.520, \beta = 1.533$ $\gamma = 1.584, \delta = 0.064$	$2V_z = 55°$ $Y = b, X \wedge c = 41°$
No. 118 Polyhalite $K_2SO_4 \cdot MgSO_4 \cdot 2CaSO_4 \cdot 2H_2O$	Triclinic. Distinct $\{100\}$ cleavage	tasteless	$\alpha = 1.547, \beta = 1.560$ $\gamma = 1.567, \delta = 0.020$	$2V_x = 64°$ orientation uncertain

Note: Gypsum, anhydrite, and the carbonates are common in evaporites.

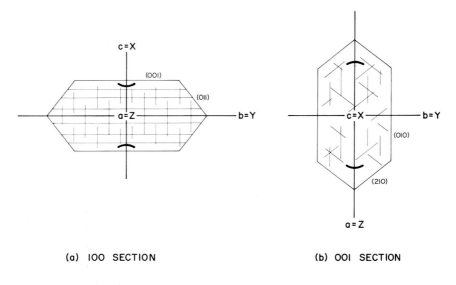

(a) 100 SECTION   (b) 001 SECTION

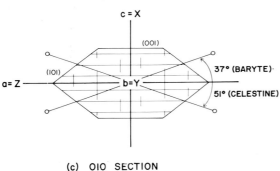

(c) 010 SECTION

Fig. 128. Orientation diagrams for baryte and celestine.

(celestine); two good cleavages at $102°$ to each other with symmetrical extinction and the O.A.P. parallel to a third cleavage.

(010) section (Fig. 128c): flash figure; δ (low) = 0.012 (baryte) or 0.009 (celestine); straight extinction with the slow direction parallel to the well-developed {001} cleavage; an oblique cross-cleavage may be visible.

Occurrence. Baryte is most commonly found in hydrothermal veins, often with Pb and Zn minerals; also as concretions in sediments, especially limestones, and more rarely as a detrital mineral or as a cement in sandstones. Celestine is found in evaporites, and in veins and fissures in carbonate sediments; also in some metalliferous veins.

Distinguishing features. The cleavages, moderate relief, low δ and occurrence are distinctive. Baryte is not easy to distinguish from celestine; an accurate measurement of R.I. or $2V$ is necessary.

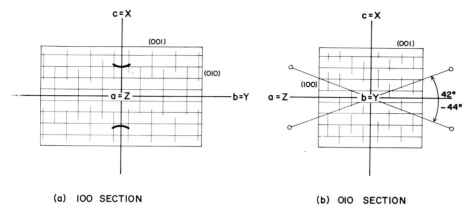

(a) 100 SECTION   (b) 010 SECTION

Fig. 129. Orientation diagrams for anhydrite.

No. 120. ANHYDRITE (+ve)   CaSO₄

Orthorhombic. Perfect {001} and good {010} and {100} cleavages which produce cube-shaped cleavage fragments. Usually massive anhedral, granular; sometimes fibrous. Simple or lamellar twinning on {010} ; if two sets of twins are developed they intersect at ca. 96°.

Colour in thin section: colourless.

Optical properties: biaxial +ve. $2V_z = 42°-44°$.
$\alpha = 1.569-1.574$, $\beta = 1.574-1.579$, $\gamma = 1.609-1.618$.
$\delta = 0.040-0.047$. $X = c$, $Y = b$, $Z = a$.

Orientation diagrams. *(100) section* (Fig. 129a): acute bisectrix figure; δ' (low) = 0.005; two good cleavages at 90° and straight extinction.
(010) section (Fig. 129b): flash figure; δ (high) = 0.040−0.047; two cleavages at 90°, straight extinction with the slow direction parallel to the best cleavage.
The (001) section displays two cleavages at 90°, provides an obtuse bisectrix figure and has a high δ' (ca. 0.035).

Occurrence. One of the prime constituents of evaporite deposits, often with gypsum and halite. It may hydrate to gypsum or form by the dehydration of gypsum. It also occurs in sediments and veins as a result of the weathering of sulphides. Also found occasionally in amygdales, and in fumarole deposits.

Distinguishing features. The occurrence, pseudocubic cleavages, and high δ are distinctive. Gypsum, baryte and celestine all have a much lower δ than anhydrite.

No. 121. GYPSUM (+ve)   CaSO₄ · 2H₂O

Monoclinic, $\beta = 114°$. Perfect {010} and distinct {100} and {011} cleavages. Euhedral crystals are {010} tablets with well developed {011} and {Ī11} forms; often massive, subhedral to anhedral, granular; also fibrous parallel to c. Twinning on {100}; lamellar twinning may be produced during thin-section making.

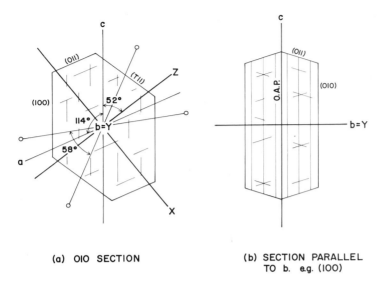

(a) OIO SECTION

(b) SECTION PARALLEL
TO b. e.g. (IOO)

Fig. 130. Orientation diagrams for gypsum.

Colour in thin section: colourless.

Optical properties: biaxial +ve. $2V_z = 58°$.
$\alpha = 1.519–1.521$, $\beta = 1.522–1.526$, $\gamma = 1.529–1.531$.
$\delta = 0.010$. $X \wedge c = 38°$, $Y = b$, $Z \wedge a = 14°$.
The $2V$ decreases with increase in temperature so that at $91°$ C, $2V_z = 0°$.
Gypsum may partially dehydrate during thin-section making to form the fibrous hemi-
hydrate $CaSO_4 \cdot \frac{1}{2}H_2O$ (*plaster of paris*). The hemihydrate has $\alpha = 1.559$ and $\gamma = 1.583$,
i.e. an R.I. and δ intermediate between those of gypsum and anhydrite; it readily reab-
sorbs water to form gypsum.

Orientation diagrams. *(010) section and cleavage fragments* (Fig. 130a): flash figure; δ
(low) = 0.010; two distinct cleavages at $114°$ to each other; inclined extinction with $Z \wedge$
$\{100\}$ cleavage traces = $52°$.
Sections parallel to b (Fig. 130b): O.A.P. is parallel to the perfect $\{010\}$ cleavage; acute
or obtuse bisectrix or optic-axis figure, centred or off-centred; may be two oblique cross
cleavages $\{011\}$ or a single cross-cleavage $\{100\}$ at $90°$ to $\{010\}$; δ' is low or very low.

Occurrence. One of the prime constituents of evaporite deposits, often with anhydrite
and halite. It may dehydrate to anhydrite or form by hydration of anhydrite. Also found
in muddy and calcareous sediments, veins, fumarole deposits, and may be produced by
the weathering of sulphides such as pyrite.

Distinguishing features. The low relief and δ, perfect cleavage, and occurrence are distinctive. The optical properties are similar to those of some zeolites, e.g. thomsonite, but
their orientations and parageneses differ.

No. 122. ALUNITE (+ve) $\qquad$ $KAl_3(SO_4)_2(OH)_6$

Trigonal. Good $\{0001\}$ cleavage. Euhedral crystals may be pseudocubic rhombohedra or $\{0001\}$ plates; often anhedral aggregates.

Colour in thin section: colourless.

Optical properties: uniaxial +ve.
$\omega = 1.572$, $\epsilon = 1.592$.
$\delta = 0.020$.
Plate-like crystals and cleavage traces are length fast with straight extinction.

Occurrence. Found in hydrothermally altered acid volcanic rocks; also in veins.

Distinguishing features: The moderate relief and δ, uniaxial +ve character and cleavage are distinctive, except for brucite which is very similar. However, the occurrences of brucite and alunite are different. Muscovite superficially resembles alunite but is length slow.

(f) Carbonates

No. 123. ARAGONITE (−ve) $\qquad$ $CaCO_3$

May contain Sr.

Orthorhombic. Imperfect $\{010\}$ cleavage. Euhedral crystals are acicular with steep pyramidal terminations, or $\{010\}$ tablets with six-sided cross-sections; also subhedral in columnar or fibrous aggregates, or anhedral. Lamellar twinning with $\{110\}$ twin planes; also pseudohexagonal cyclic twinning on $\{110\}$. Effervesces like calcite in dilute HCl.

Colour in thin section: colourless.

Optical properties: biaxial −ve. $2V_x = 18°$.
$\alpha = 1.530$–1.531, $\beta = 1.680$–1.682, $\gamma = 1.685$–1.686.
$\delta = 0.155$. $X = c$, $Y = a$, $Z = b$.

Orientation diagrams. *(001) section* (Fig. 131a): acute bisectrix figure; δ' (low or very low) < 0.005; poor single cleavage; extinction may be indistinct due to proximity of optic axes; intersecting sets of twins on $\{110\}$ may be visible.
(100) section (Fig. 131b): flash figure; δ (very high) $= 0.155$; single poor cleavage, straight extinction and length fast; twin planes are very oblique to this section.
(010) section (Fig. 131c): obtuse bisectrix (flash) figure; δ' (very high) $= 0.150$; no cleavage, straight extinction and length fast in prismatic crystals; $\{110\}$ twins parallel to c may be visible.

Occurrence. Aragonite is metastable at ordinary temperatures and pressures. Nevertheless, many calcareous shells are built of aragonite, sometimes together with calcite as in some bivalves, and aragonite is an important constituent of modern shallow-water calcareous sediments. It readily changes to calcite so that it is not found in most ancient limestones. Aragonite also occurs in vesicles and cavities in volcanic rocks, especially basic volcanics, and as a high-pressure modification of calcite in regionally metamorphosed rocks of the lawsonite—glaucophane schist facies.

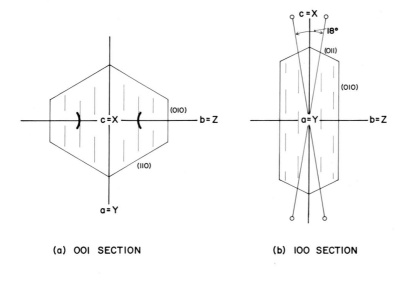

(a) OOI SECTION   (b) IOO SECTION

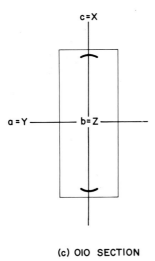

(c) OIO SECTION

Fig. 131. Orientation diagrams for aragonite.

Distinguishing features. The very high δ and effervescence in dilute acid may make distinction between aragonite and calcite difficult, especially in modern fine-grained carbonate sediments. Aragonite is distinguished by its single poor cleavage and small $2V$, but in fine-grained material, stain tests (referred to under calcite) may be necessary.

No. 124. THE TRIGONAL (RHOMBOHEDRAL) CARBONATE GROUP

No. 124A. Calcite (—ve)	$CaCO_3$
No. 124B. Dolomite (—ve)	$CaMg(CO_3)_2$
No. 124C. Ankerite (—ve)	$Ca(Mg,Fe^{2+},Mn)(CO_3)_2$
No. 124D. Magnesite (—ve)	$MgCO_3$
No. 124E. Siderite (—ve)	$FeCO_3$
No. 124F. Rhodochrosite (—ve)	$MnCO_3$

Chemical composition. The following solid-solution series are possible at low temperatures:

(1) Up to approximately 50% Mn may substitute for Ca to form Mn-rich calcite, but there is an immiscibility gap between this and rhodochrosite.

(2) Only small amounts of Mg may substitute for Ca in calcite. High-magnesium calcites in modern sediments commonly contain up to about 10 mol.% $MgCO_3$, but these are unstable, and calcite in ancient limestones is usually close to pure $CaCO_3$.

(3) Dolomite is a true double salt with equal proportions of Ca and Mg. There is no solid solution series between dolomite and either calcite or magnesite. There is complete solid solution with ankerite in which there is substantial replacement of Mg by Fe^{2+} and Mn.

(4) There are solid solution series between magnesite and siderite and between siderite and rhodochrosite.

Crystallography. Trigonal. Perfect $\{10\bar{1}1\}$ cleavages (rhombohedral). *Calcite* may be euhedral, commonly as elongate prisms with scalenohedral or rhombohedral terminations, or as rhombohedra; also anhedral, massive, stalactitic, etc. *Dolomite, ankerite,* and *siderite* may form euhedral rhombohedra; also massive, anhedral. *Magnesite* and *rhodochrosite* are usually massive, granular. Growth twins, especially on $\{0001\}$ and $\{10\bar{1}1\}$ are common in calcite, dolomite, and ankerite. Lamellar twins produced by deformation are very common on $\{01\bar{1}2\}$ in calcite (Fig. 133); lamellar glide-twins on $\{02\bar{2}1\}$ are produced in dolomite only at temperatures > ca. 300° C. Twinning is rare in rhodochrosite and siderite, and absent in magnesite. A short discussion of glide mechanisms in calcite and dolomite is given after the section on distinguishing features.

Colour in thin section: usually colourless; siderite may be pale yellow-brown and rhodochrosite pale pink.

Optical properties: uniaxial —ve.
R.I. and δ values for pure end-members are:

Calcite: $\omega = 1.658$, $\epsilon = 1.486$, $\delta = 0.172$.
Dolomite: $\omega = 1.679$, $\epsilon = 1.500$, $\delta = 0.179$.
Magnesite: $\omega = 1.700$, $\epsilon = 1.509$, $\delta = 0.191$.
Siderite: $\omega = 1.875$, $\epsilon = 1.635$, $\delta = 0.242$.
Rhodochrosite: $\omega = 1.816$, $\epsilon = 1.597$, $\delta = 0.219$.
The variation in the value of ω with chemistry is given in Fig. 132 (ω can be measured in any grain). Since crushed material always lies on the perfect $\{10\bar{1}1\}$ cleavages, ϵ is difficult to measure. However, the value of ϵ can be determined from·ϵ' measured on cleavage fragments with the use of a chart provided by Loupekine (1947).

Occurrence. *Calcite* is found as a primary or secondary mineral in a very wide range of rock types. It is the prime constituent of most limestones as a primary precipitate, as a

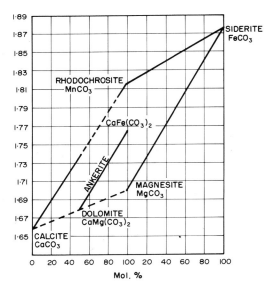

Fig. 132. Variation of ω with composition in the trigonal carbonates. Modified with permission from Kennedy (1947, fig. 8).

mineral in shell fragments, and as a diagenetic mineral. Metastable aragonite and high-Mg calcite are common in modern shallow-sea limestones but eventually change to calcite, the Mg being flushed out or sometimes forming dolomite. Calcite occurs as ooliths in some limestones. Calcite is a common cement in other sediments, and is precipitated from groundwaters to form travertine deposits and stalactites, etc. It is found in many metamorphic rocks of low to high grade, particularly metamorphosed limestones (marbles). In some circumstances, CO_2 may be lost during metamorphism, and the Ca combines with silica to form minerals such as wollastonite, diopside, tremolite, grossularite, vesuvianite, etc. In igneous rocks, calcite is usually secondary, occurring in vesicles, fissures, and replacing a wide range of minerals, especially plagioclase; it also occurs as a primary magmatic mineral in carbonatites and associated rocks. Calcite is common in hydrothermal veins.

Dolomite is most commonly found in sediments, often with calcite; some is primary, but much is secondary. Secondary dolomite is often prominent as euhedral rhombohedra. Low-grade metamorphism of dolomite sediments produces dolomitic marble, but at high grades the dolomite may dissociate to calcite and periclase (often altered to brucite). Dolomite may react with silica during metamorphism to form minerals such as talc and tremolite. Dolomite also occurs as a primary mineral in some carbonatites, and as a hydrothermal mineral in veins.

Ankerite is most common as a secondary mineral in Fe-rich sediments and coal, especially in veins. It is also found as a primary magmatic mineral in carbonatites, and in metamorphosed Fe-rich sediments.

Magnesite forms as an alteration product of Mg-rich igneous and metamorphic rocks. Talc-magnesite schists are produced by the low-temperature hydrothermal metamorphism of serpentinite. Magnesite also occurs in some evaporites.

Siderite is a secondary mineral in some Fe-rich sediments; it also occurs in hydrothermal veins, and in metamorphosed Fe-rich sediments.

Fig. 133. Typical deformation twinning in calcite. Mt. Arthur Marble, New Zealand (view measures 3.1 mm × 2.0 mm). Crossed-polarised light.

Rhodochrosite is rare and found in association with other Mn-rich minerals in ore deposits, veins and sediments. It alters to black opaque pyrolusite (MnO_2).

Distinguishing features. The very high δ and the perfect rhombohedral cleavages are distinctive of the group. The "twinkling" of calcite, dolomite, ankerite and magnesite is characteristic, and serves to distinguish them from other minerals with a very high δ such as sphene. Distinction amongst the trigonal carbonates is less easy. Measurement of ω (Fig. 132) is not always sufficient to identify a particular species. Dolomite, ankerite and siderite are characterised by their common euhedral (rhombohedral) habit. In deformed rocks, calcite is characterised by abundant twin-lamellae (Fig. 133); dolomite does not develop glide-twins below 300°C, and magnesite always lacks glide-twinning. The twin-lamellae of calcite are $\{01\bar{1}2\}$ and those of dolomite $\{02\bar{2}1\}$; in sections displaying two intersecting sets of twins, ω (the direction of higher relief) lies in the acute angle between the lamellae in calcite but in the obtuse angle in dolomite. Calcite effervesces readily in cold dilute HCl whereas dolomite does not.

Stain tests. In order to make modal and textural analyses of carbonate sediments, it may be necessary to distinguish the various carbonate minerals using stain tests. Numerous methods have been devised, and they are discussed fully in Wolf et al. (1967). Some experience is necessary in their use since the effects of staining vary with differing grain size and porosity.

Glide mechanisms in calcite and dolomite. Calcite deforms very readily by translation and twin-gliding. Extensive laboratory investigations (summarised in Turner and Weiss, 1963) have shown that the principal glide mechanisms are: twin-gliding on $\{01\bar{1}2\}$ effective under all conditions of temperature and pressure investigated; translation on $\{10\bar{1}1\}$ especially at temperatures up to 400° C; translation on $\{02\bar{2}1\}$ especially at temperatures >500° C. Twinning generally occurs in preference to translation if the sense of shear is suitable. Gliding causes rotation, bending and flattening of grains; complex patterns of glide-twins are commonly observed in thin sections (Fig. 133). The strain caused by gliding may assist nucleation of new grains which grow at the expense of earlier-strained crystals (process of recrystallisation). A preferred orientation of *c*-axes develops during deformation and recrystallisation.

Dolomite deforms less easily, but suffers translation gliding on $\{0001\}$ at low temperatures if suitably oriented in the stress field. Dolomite is unusual in that it becomes stronger with increasing temperature; twin-gliding on $\{02\bar{2}1\}$ does not occur below 300° C and becomes more important than translation at high temperatures.

(g) Phosphates

No. 125. APATITE (−ve) $Ca_5(PO_4)_3(OH,F,Cl)$

C replaces some of the P in the related minerals *dahllite* (carbonate-apatite with F < 1%) and *francolite* (carbonate-apatite with F > 1%). *Collophane* is a term used for cryptocrystalline dahllite or francolite.

Hexagonal. No good cleavage. Euhedral crystals of apatite are short, long or acicular prisms with a hexagonal cross-section; also subhedral or anhedral grains. Dahllite is the inorganic phase in bone, and dahllite and francolite form concretions and fibrous radiating or spherulitic aggregates.

Colour in thin section: usually colourless; dahllite, francolite and collophane are commonly yellow-brown or grey.

Optical properties. *Apatite*: uniaxial −ve; occasionally biaxial with a small 2*V*.
ω = 1.632—1.668, ϵ = 1.628—1.665.
δ = 0.002—0.008.
The R.I. of chlorapatite > hydroxyapatite > fluorapatite.
Related minerals: the R.I. of dahllite varies in the range 1.52—1.61 (usually 1.55—1.59), and that of francolite is <1.63. The δ of dahllite and francolite varies from very low to as high as 0.016, and they may be biaxial. Collophane is isotropic.

Occurrence. Apatite is a very common accessory mineral in almost all types of igneous and metamorphic rocks, normally as very small grains or prisms; it may be abundant in some alkaline types and carbonatites. It is found as a detrital mineral in many sediments.

The inorganic phase in bone material is dahllite, and dahllite, francolite or collophane are prime constituents in rock phosphate, secondary phosphatic material in sedimentary rocks, and coprolites.

Distinguishing features. The moderate relief, low or very low δ, —ve character, and lack of cleavage or colour in apatite are distinctive. The brown colour, occurrence, and amorphous, fibrous or skeletal character of collophane, francolite and dahllite are distinctive.

No. 126. MONAZITE (+ve)   (Ce,La)PO$_4$

Th, Nd and other rare elements present.

Monoclinic, β = 104°. Distinct {100} and {001} cleavages. Usually small crystals elongate parallel to b or {100} tablets. Twinning with {100} twin planes common.

Colour in thin section: colourless or pale yellow.

Colour in detrital grains: yellow, brown, or red; pleochroism weak with $Y > X = Z$.

Optical properties: biaxial +ve. $2V_z$ = 3°—19°.
α = 1.770—1.800, β = 1.777—1.801, γ = 1.825—1.850.
δ = 0.045—0.075. $X = b$, $Y \wedge a$ = 7°—12°, $Z \wedge c$ = 2°—7°.
{100} tablets have slightly inclined or parallel extinction. Crystals elongate parallel to b are length fast with straight extinction. {001} cleavage fragments provide an almost centred acute bisectrix figure.

Occurrence. A rare accessory mineral in acid and alkaline igneous rocks and gneisses. May cause pleochroic haloes in biotite. It is resistant to weathering and may be concentrated as a detrital mineral in sufficient quantities to be economically important as a source of Ce and other rare elements.

Distinguishing features. The very high relief, high δ, and colour are characteristic. Detrital grains are usually well rounded or egg-shaped. It may be confused with zircon, sphene and xenotime, but zircon has a higher R.I. and is uniaxial, and both sphene and xenotime have a higher δ.

No. 127. XENOTIME (+ve)   YPO$_4$

Other rare-earths are usually present.

Tetragonal. Distinct {110} cleavages. Crystals are prismatic with pyramidal terminations, and resemble zircon.

Colour in thin section: colourless, yellow or brown.

Colour in detrital grains: generally yellow or brown; weak pleochroism.

Optical properties: uniaxial +ve.
ω = 1.720—1.724, ϵ = 1.810—1.828.
δ = 0.086—0.107.
Prismatic crystals are length slow with straight extinction.

Occurrence. A rare accessory mineral in acid and alkaline igneous rocks and gneisses, often associated with zircon. Xenotime may cause pleochroic haloes in biotite. Commonly found as detrital grains in heavy-mineral suites.

Distinguishing features. Often mistaken for zircon, but xenotime is distinguished by its lower R.I. and higher δ.

References

Albee, A.L., 1962. Relationships between the mineral association, chemical composition and physical properties of the chlorite series. *Am. Mineral.*, 47: 851—870.

Ashworth, J.R., 1972. Myrmekites of exsolution and replacement origins. *Geol. Mag.*, 109: 45—62.

Bailey, E.H. and Stevens, R.E., 1960. Selective staining of plagioclase and K-feldspar on rock slabs and thin sections. *Geol. Soc. Am. Bull.*, 71: 2047.

Barker, D.S., 1970. Compositions of granophyre, myrmekite, and graphic granite. *Geol. Soc. Am. Bull.*, 81: 3339—3350.

Barth, T.F.W., 1969. *Feldspars*. Wiley-Interscience, New York, N.Y., 261 pp.

Bennett, R.L., 1958. Evaporite sections. *J. Inst. Sci. Technol.*, 4: 358.

Borchert, H. and Muir, R.O., 1964. *Salt Deposits*. D. van Nostrand, London, 338 pp.

Bottinga, Y., Kudo, A. and Weill, D., 1966. Some observations on oscillatory zoning and crystallization of magmatic plagioclase. *Am. Mineral.*, 51: 792—806.

Bowen, N.L., 1913. The melting phenomena of the plagioclase feldspars. *Am. J. Sci.*, 35: 577—599.

Bowen, N.L. and Tuttle, O.F., 1950. The system $NaAlSi_3O_8—KAlSi_3O_8—H_2O$. *J. Geol.*, 58: 489—511.

Boyd, F.R. and Schairer, J.F., 1964. The system $MgSiO_3—CaMgSi_2O_6$. *J. Petrol.*, 5: 275—309.

Brown, G.M. and Vincent, E.A., 1963. Pyroxenes from the late stages of fractionation of the Skaergaard Intrusion, East Greenland. *J. Petrol.*, 4: 175—197.

Coombs, D.S., 1971. Present status of the zeolite facies. In: *Molecular Sieve Zeolites, I. Advances in Chemistry Series No. 101*. American Chemical Society, Washington, D.C., pp. 317—327.

Dana, E.S., 1932. *A Textbook of Mineralogy*, Wiley and Sons, New York, N.Y., 4th ed. (by W.E. Ford), 851 pp.

Deer, W.A., Howie, R.A. and Zussman, J., 1962. *Rock-Forming Minerals, 1, 3 and 5*. Longmans, London, Vol. 1: 333 pp., Vol. 3: 270 pp., Vol. 5: 371 pp.

Deer, W.A., Howie, R.A. and Zussman, J., 1963. *Rock-Forming Minerals, 2 and 4*. Longmans, London, Vol. 2: 379 pp., Vol. 4: 435 pp.

Deer, W.A., Howie, R.A. and Zussman, J., 1966. *An Introduction to the Rock-Forming Minerals*. Longmans, London, 528 pp.

Emmons, R.C., 1943. The universal-stage. *Geol. Soc. Am. Mem.*, 8: 205 pp.

Ernst, W.G., 1968. *Minerals, Rocks and Inorganic Materials, 1. Amphiboles*. Springer-Verlag, New York, N.Y., 125 pp.

Fairbairn, H.W., 1949. *Structural Petrology of Deformed Rocks*. Addison-Wesley, Cambridge, Mass., 344 pp.

Flinn, D., 1973. Two flow-charts of orthoscopic U-stage techniques. *Mineral. Mag.*, 39: 368—370.

Fraser, W.E. and Downie, G., 1964. The spectrochemical determination of feldspars within the field microcline-albite-labradorite. *Mineral. Mag.*, 33: 790—798.

Frondel, C., 1962. *Dana's System of Mineralogy, III. Silica Minerals*. Wiley and Sons, New York, N.Y., 334 pp.

Galopin, R. and Henry, N.F.M., 1972. *Microscopic Study of Opaque Minerals*. Heffer and Sons, Cambridge, 322 pp.

Gomes, C. de B., Moro, S.L. and Dutra, C.V., 1970. Pyroxenes from the alkaline rocks of Itapirapua, Sao Paulo, Brazil. *Am. Mineral.*, 55: 224—230.

Gregnanin, A. and Viterbo, C., 1965. Metodo di colorazione per identificare la cordierite in sezione sottile. *Rend. Soc. Mineral. Ital.*, 21: 111—120.

Grim, R.E., 1968. *Clay Mineralogy*. McGraw-Hill, New York, N.Y., 2nd. ed., 596 pp.

Harrington, V.F. and Buerger, M.J., 1931. Immersion liquids of low refraction. *Am. Mineral.*, 16: 45—54.

Hess, H.H., 1949. Chemical composition and optical properties of common clino-pyroxenes, I. *Am. Mineral.*, 34: 621—666.

Hurlbut, C.S., 1971. *Dana's Manual of Mineralogy*. Wiley and Sons, New York, N.Y., 18th ed., 579 pp.

Jones, J.B., Sanders, J.V. and Segnit, E.R., 1964. Structure of opal. *Nature*, 204: 990—991.

Kennedy, G.C., 1947. Charts for correlation of optical properties with chemical composition of some common rock-forming minerals. *Am. Mineral.*, 32: 561—573.

Kuno, H., 1955. Ion substitution in the diopside—ferropigeonite series of clinopyroxenes. *Am. Mineral.*, 40: 70—93.

Laduron, D.M., 1971. A staining method for distinguishing paragonite from muscovite in thin section. *Am. Mineral.*, 56: 1117—1119.

Lancelot, Y., 1973. Chert and silica diagenesis in sediments from the Central Pacific. In: E.L. Winterer, J.L. Ewing et al., *Initial reports of the Deep Sea Drilling Project, XVII.* U.S. Government Printing Office, Washington, D.C., pp. 377—405.

Laniz, R.P., Stevens, R.E. and Norman, M.B., 1964. Staining of plagioclase feldspar and other minerals with F.D. and C. Red. No. 2. *U.S. Geol. Survey Prof. Paper*, 501-B: 152—153.

Larsen, E.S. and Berman, H., 1934. The microscopic determination of the non-opaque minerals. *U.S. Geol. Survey Bull.*, 848: 266 pp.

Leake, B.E., 1968. Optical properties and composition in the orthopyroxene series. *Mineral. Mag.*, 36: 745—747.

Loupekine, I.S., 1947. Graphical derivation of refractive index ε for the trigonal carbo-nates. *Am. Mineral.*, 32: 502—507.

Lyons, P.C., 1971. Staining of feldspars on rock-slab surfaces for modal analysis. *Mineral. Mag.*, 38: 518—519.

MacKenzie, W.S. and Smith, J.V., 1956. The alkali-feldspars, III. An optical and X-ray study of high-temperature feldspars. *Am. Mineral.*, 41: 405—427.

Mathews, W.H., 1951. A useful method for determining approximate composition of fine-grained igneous rocks. *Am. Mineral.*, 36: 92—101.

Meyrowitz, R., 1955. A compilation and classification of immersion media of high index of refraction. *Am. Mineral.*, 40: 398—409.

Morse, S.A., 1970. Alkali feldspars with water at 5 kb pressure. *J. Petrol.*, 11: 221—251.

Muir, I.D., 1951. The clinopyroxenes of the Skaergaard intrusion, eastern Greenland. *Mineral. Mag.*, 29: 690—714.

Myer, G.H., 1966. New data on zoisite and epidote. *Am. J. Sci.*, 264: 364—385.

Nakamura, Y. and Kushiro, I., 1970. Equilibrium relations of hypersthene, pigeonite and augite in crystallising magmas: microprobe study of a pigeonite andesite from Weisel-berg, Germany. *Am. Mineral.*, 55: 1999—2015.

Oehler, J.H., 1973. Tridymite-like crystals in cristobalitic "cherts". *Nature Phys. Sci.*, 241: 64—65.

Phillips, W.R., 1964. A numerical system of classification for chlorites and septechlorites. *Mineral. Mag.*, 33: 1114—1124.

Phillips, F.C., 1971. *An Introduction to Crystallography*. Oliver and Boyd, Edinburgh, 4th ed., 351 pp.

Poldervaart, A., 1950. Correlation of physical properties and chemical composition in the plagioclase, olivine, and orthopyroxene series. *Am. Mineral.*, 35: 1067—1079.

Poldervaart, A. and Hess, H.H., 1951. Pyroxenes in the crystallisation of basaltic magma. *J. Geol.*, 59: 472—489.

Reed, F.S. and Mergner, J.L., 1953. Preparation of rock thin sections. *Am. Mineral.*, 38: 1184—1203.

Shand, S.J., 1939. On the staining of feldspathoids, and on zonal structure in nepheline. *Am. Mineral.*, 24: 508—513.

Shelley, D., 1971. Hypothesis to explain the preferred orientations of quartz and calcite produced during syntectonic recrystallisation. *Geol. Soc. Am. Bull.*, 82: 1943—1954.

Shelley, D., 1972. Syntectonic recrystallisation and preferred orientation of quartz. *Geol. Soc. Am. Bull.*, 83: 3523—3524.

Shelley, D., 1973. Myrmekites from the Haast Schists, New Zealand. *Am. Mineral.*, 58: 332—338.

Slemmons, D.B., 1962. Determination of volcanic and plutonic plagioclases using a three or four-axis universal stage. *Geol. Soc. Am. Spec. Paper*, 69: 64 pp.

Smith, J.R., 1958. The optical properties of heated plagioclases. *Am. Mineral.*, 43: 1179—1194.

Sriramadas, A., 1957. Diagrams for the correlation of unit cell edges and refractive indices with the chemical composition of garnets. *Am. Mineral.*, 42: 294—298.

Starkey, J., 1967. On the relationship of periclase and albite twinning to the composition and structural state of plagioclase feldspars. *Schweiz. Mineral. Petrogr. Mitt.*, 47: 257—268.

Taylor, J.C.M., 1960. Impregnation of rocks for sectioning. *Geol. Mag.*, 97: 261.

Tobi, A.C., 1956. A chart for measurement of optic axial angles. *Am. Mineral.*, 41: 516—519.

Tobi, A.C., 1963. Plagioclase determination with the aid of the extinction angles in sections normal to (010). A critical comparison of current albite-Carlsbad charts. *Am. J. Sci.*, 261: 157—167.

Tobi, A.C. and Kroll, H., manuscript in preparation.

Tröger, W.E., 1971. *Optische Bestimmung der gesteinsbildenden Minerale, 1. Bestimmungstabellen*. E. Schweizbart'sche Verlagsbuchhandlung, Stuttgart, 4th ed. (by H.U. Bambauer, F. Taborszky and H.D. Trochim), 188 pp.

Tullis, J., Christie, J.M. and Griggs, D.T., 1973. Microstructures and preferred orientations of experimentally deformed quartzites. *Geol. Soc. Am. Bull.*, 84: 297—314.

Tunell, G., 1952. The angle between the a-axis and the trace of the rhombic section on the {010} pinacoid in the plagioclases. *Am. J. Sci.*, Bowen Volume, pp. 547—551.

Turner, F.J. and Verhoogen, J., 1960. *Igneous and Metamorphic Petrology*. McGraw-Hill, New York, N.Y., 2nd ed., 694 pp.

Turner, F.J. and Weiss, L.E., 1963. *Structural Analysis of Metamorphic Tectonites*. McGraw-Hill, New York, N.Y., 545 pp.

Tuttle, O.F., 1952. Optical studies on alkali-feldspars. *Am. J. Sci.*, Bowen Volume, pp. 553—567.

Tuttle, O.F. and Bowen, N.L., 1958. Origin of granite in the light of experimental studies in the system $NaAlSi_3O_8—KAlSi_3O_8—SiO_2—H_2O$. *Geol. Soc. Am. Mem.*, 74: 153 pp.

Vance, J.A., 1961. Polysynthetic twinning in plagioclase. *Am. Mineral.*, 46: 1097—1119.

Vance, J.A., 1969. On synneusis. *Contrib. Mineral. Petrol.*, 24: 7—29.

Von Burri, C., Parker, R.L. and Wenk, E., 1967. *Die optische Orientierung der Plagioklase*. Birkhäuser Verlag, Basel, 334 pp.

Wahlstrom, E.E., 1969. *Optical Crystallography*. Wiley and Sons, New York, N.Y., 489 pp.

Walker, G.P.L., 1960. Zeolite zones and dike distribution in relation to the structure of the basalts of eastern Iceland. *J. Geol.*, 68: 515—528.

Weaver, C.F. and McVay, T.N., 1960. Immersion oils with indices of refraction from 1.292 to 1.411. *Am. Mineral.*, 45: 469—470.

Winchell, A.N., 1939. *Elements of Optical Mineralogy, III. Determinative Tables*. Wiley and Sons, New York, N.Y., 2nd ed., 231 pp.

Winchell, A.N., 1951. *Elements of Optical Mineralogy, II. Description of Minerals*. Wiley and Sons, New York, N.Y., 4th ed. (in collaboration with H. Winchell), 551 pp.

Winchell, H., 1958. The composition and physical properties of garnet. *Am. Mineral.*, 43: 595—600.

Winchell, H., 1965. *Optical Properties of Minerals: A Determinative Table*. Academic Press, New York, N.Y., 91 pp.

Wolf, K.H., Easton, A.J. and Warne, S., 1967. Techniques of examining and analysing carbonate skeletons, minerals, and rocks. In: G.V. Chilingar, H.J. Bissell and R.W. Fairbridge (Editors), *Carbonate Rocks. Developments in Sedimentology, 9B*. Elsevier, Amsterdam, pp. 253—342.

Wright, H.G., 1964. The use of epoxy resins in the preparation of petrographic thin sections. *Mineral. Mag.*, 33: 931—933.

Zeck, H.P., 1972. Irrational composition planes in cordierite sector trillings. *Nature Phys. Sci.*, 238: 47—48.

Index

Note: the main reference to each mineral is printed in boldface, and the consecutive numbering system is given in brackets.